Science and Human Freedom

Michael Esfeld

Science and Human Freedom

palgrave
macmillan

Michael Esfeld
Section de Philosophie
University of Lausanne
Lausanne, Switzerland

ISBN 978-3-030-37773-1   ISBN 978-3-030-37771-7   (eBook)
https://doi.org/10.1007/978-3-030-37771-7

This Palgrave Macmillan imprint is published by the registered company Springer Nature Switzerland AG.
The registered company address is: Gewerbestrasse 11, 6330 Cham, Switzerland

Introduction

The age of enlightenment has two faces. On the one hand, there is the liberation of humankind as expressed, for instance, in Immanuel Kant's (1784) definition of enlightenment as "man's emergence from his self-imposed immaturity".[1] On the other hand, there is scientism, that is, the idea that scientific knowledge is unlimited, encompassing also humankind and all aspects of our existence, as illustrated, for instance, in Julien Offray de La Mettrie's *L'homme machine* (1747). Both have the rejection of knowledge claims of traditional authorities (such as the church) in common. But whereas the former is about giving each person the freedom to take their own decisions, the latter paves the way for assuming that scientific knowledge is in the position to predetermine the appropriate decisions, both individually and collectively.

These two faces of enlightenment can be traced back to antiquity. According to Aristotle's *Politics*, the organization of society is a matter of decisions that the citizens have to take in common deliberation. It is not predetermined by any knowledge. For Plato, by contrast, it is a question of knowledge how to conduct one's individual life as well as society. Consequently, as he sets out in the *Republic*, the philosophers should rule. In modern times, scientific knowledge then takes the place of the

---

[1] Quoted from Kant (1983, p. 41); "der Ausgang des Menschen aus seiner selbst verschuldeten Unmündigkeit" in the German original.

knowledge of which Plato assumes that it can be acquired by philosophical contemplation.

Accordingly, this book is in the first place about what the scientific image of the world is and what are its limits. Its central objective is to bring out how science makes us free and thereby contributes to the open society (in the sense of Popper's famous book *The open society and its enemies* (1945), which was the first philosophy book that I read). The present book therefore first works out why natural science, through the laws that it discovers, strengthens our freedom instead of infringing upon it; building on this, I will then show why it is wrong-headed to assume that science can give us the norms to design society and our individual lives. This mistake originates in enlightenment personalities such as La Mettrie, it is later implemented in Marxism and it is fuelled today by a misapprehension of the discoveries in physics, evolutionary biology, genetics, neuro- and cognitive science, etc. Giving science such an unjustified power, then, provokes the reaction to refuse recognizing that science discovers truth about the world. Unfortunately, this reaction is also widespread among postmodernist intellectuals. It invites abandoning the demarcation line between fact and fake. It thereby jettisons not only scientism, but also the idea of science contributing to the liberation of humankind.

Consequently, this book shows what is wrong with the widespread claims to the effect that scientific laws (such as, notably, universal and deterministic laws in physics), scientific discoveries (such as, for instance, discoveries in genetics or cognitive science) and scientific explanations (such as, for instance, explanations of human behaviour in evolutionary biology or neuroscience) infringe upon human free will. In brief, in the first place, the ontology of science—that is, what has to be admitted as existing in the world in order to make the truth of scientific theories intelligible—is not rich enough to entitle conclusions to that effect. Moreover, scientific laws, discoveries and explanations are about contingent facts by contrast to something that is necessary (that is, that could not have been otherwise). Most importantly, scientific theories are conceived, endorsed and justified in normative attitudes of giving and asking for reasons that presuppose the freedom of persons in formulating, testing and judging theories. For that reason, persons cannot be subsumed under the scientific image of the world. Hence, science gives us information about the world that can serve as guide for our actions, but not norms, neither for the individual life, nor for society—on pain of committing what is known as the naturalistic fallacy (that is, the attempt to deduce norms from facts). Science makes us free in that it shows that we have the freedom to set up the norms for what to think and how to act—as individuals as well as in societies—, but thereby also the burden of the responsibility for our thoughts and actions.

What is science? At least the following three traits distinguish science from other human enterprises, including other intellectual ones:

- *Objectivity*: What science tells us about the world does not depend on any particular viewpoint. Science is independent of gender, race, religion, or geographical or temporal location. Scientific theories propose a point of view from nowhere and nowhen—although, of course, they have a particular origin; but their validity is independent of that origin. Everybody can become part of the scientific community. There is no Chinese mathematics, physics or biology in contrast to an American one. The same applies to philosophy insofar as it is an argumentative enterprise that strives for knowledge about the world and our position in it.
- *Systematicity*: A scientific theory seeks to represent as many phenomena as possible in terms of as simple a law as possible. Prominent examples are the law of natural selection in evolutionary biology and the law of gravitation in physics. The latter is an ideal example of a law of nature, because it applies to everything in the universe.
- *Confirmation by evidence*: Any claim in science has to be such that it can be confirmed by evidence that is accessible independently of the claim in question. That is, the claim has to allow the derivation of predictions that can be checked without presupposing the claim at issue. For instance, Einstein's theory of gravitation predicts that starlight passing by the sun will be bent by the gravitational field of the sun. This can be observed at a solar eclipse (first done in 1919). The observation of this phenomenon is independent of the theoretical claims of general relativity theory about the geometry of space and time and the behaviour of the gravitational field. As this example

shows, confirmation does not always imply intervention by means of experiments. The crucial issue is observation of new phenomena predicted by and made intelligible by the theory.

Laying stress on these features as characterizing science usually is associated with the stance that is known as *scientific realism*, signifying, in brief, that science reveals the constitution of the natural world. If a human enterprise can achieve that goal at all, only science can do so. This book is committed to scientific realism. The crucial point in our context is that laying stress on these features does not prevent us from acknowledging the limits of scientific explanations and, notably, realize how science makes us free instead of infringing upon our freedom.

In a broader perspective, this book is an essay on the interplay between what Wilfrid Sellars (1962) calls the *scientific* and the *manifest image of the world*. The manifest image is not common sense. It is the philosophically reflected view of the world that puts persons at the centre, taking them to be irreducible to something more fundamental and thus endorsing them as ontologically primitive. Consequently, one does not answer the question of the relationship between these two images by showing how one can explain the familiar macroscopic world on the basis of fundamental physics.

This book distinguishes between three ways how to conceive the relationship between these two images:

1. *The scientific image is complete*: Persons can be reduced to the ontology of science via functional definitions, on a par with everything else that does not figure explicitly in the ontology of science. In the last resort, this is the ontology of fundamental physics. That is to say: the persons that exist in the world are identical with certain specific configurations of matter and their behaviour under certain conditions in the environment. A complete physical description of the world entails also all the true propositions about persons, including the rules they follow and should follow in their thoughts and actions.
2. *The manifest image is complete*: Everything that there is in the world is in some way or other analogous to persons. Scientific theories that abstract from the features that are analogous to persons are only of

instrumental use for efficient predictions. They do not reveal the essence of the world.

3. *Each image captures only a part of what exists*: The scientific image tells the truth about the world when leaving the features that characterize persons aside. These features exist and are ontologically primitive on a par with matter in motion.

Following Kant's enlightenment philosophy and Sellars's call for a synoptic view of both these images, the book argues for a particular version of (3): the scientific image—as well as any scientific theory—presupposes the freedom of persons in forming concepts, building up and justifying theories. However, being a person is not a fact, a property or a substance in addition to the material ones. It is an attitude that one adopts to oneself and others. In adopting this attitude, one brings oneself into existence as a being that creates meaning and thereby rules for thought and action and that, consequently, has to justify what it thinks and what it does.

On this basis, the book argues for a twofold conception of freedom: in the first place, there is freedom in the sense that the laws of science, even if they are universal and deterministic laws, neither predetermine our motions, nor the motions of any other objects. First comes the motion of matter, then come the theories and the laws that reveal contingent patterns or regularities in these motions. If the scientific image were the complete image, this would be all the freedom that there is. However, if one acknowledges that the scientific image is conceived, endorsed and justified by persons in normative attitudes of giving and asking for reasons, one realizes that there is a freedom that is characteristic of persons only and that is a freedom from matter in motion. It is the freedom to set up norms for thought and action (indeed, the freedom to have to set up such norms). Again, there is nothing in science that prevents us from having our actions shaped by this freedom.

The book is organized in three parts or chapters. The next chapter works out what the ontological commitments of science are and what they are not. It focuses on the fundamental and universal theories of physics from Newtonian mechanics to today's quantum physics. The chapter answers the following question: Which ontological commitments are minimally sufficient to understand our scientific knowledge? The purpose of this chapter is not to teach physics, although it will go into some physical details. The objective is to work out the philosophical points that are necessary in order to grasp why science does not come into conflict with our freedom.

Chapter 2 then goes into the achievements as well as the limits of scientific laws and explanations. It leads to making the case for science bringing out our freedom instead of infringing upon it. By the end of Chap. 2, we will have obtained an argument to the effect that there is no basic conflict between the scientific and the manifest image as regards time and free will (both are interconnected: without openness for change and time as its measure there is no free will). Such conflicts are, to use the term of Rudolf Carnap (1928), pseudo-problems (*Scheinprobleme*) that result from a misapprehension of the ontological commitments of scientific theories.

Against this background, Chap. 3 considers the focal point of the conflict between the scientific and the manifest image of the world, namely normativity, which concerns not only human action, but already thought. The chapter then elaborates on how both images lead to human freedom, sets out the mentioned twofold conception of freedom and goes into the consequences of that freedom, pointing out that there is no knowledge—scientific or otherwise—that infringes upon freedom. The summary at the end provides an overview of the main propositions of the book.

For fruitful comments and discussions I would like to thank my collaborators and the participants of my research seminar at the University of Lausanne in the academic year 2018/19—especially Guillaume Köstner and Christian Sachse—, the collaborators of the Center for Advanced Studies "Imaginaria of force" at the University of Hamburg—especially Frank Fehrenbach and Cornelia Zumbusch for the invitation in the summer term 2019—, as well as Andreas Hüttemann, Ingvar Johansson, Barry Loewer, Anna Marmodoro, Daniel von Wachter and Gerhard Wagner.

References

Carnap, Rudolf (1928): *Scheinprobleme in der Philosophie. Das Fremdpsychische und der Realismusstreit.* Berlin-Schlachtensee: Weltkreis Verlag.

LaMettrie, Julien Offray de (1747): *L'homme machine*. Leyden.

Popper, Karl (1945): *The open society and its enemies*. London: Routledge.

Kant, Immanuel (1983): *Perpetual peace and other essays on politics, history, and morals*. Translated by Ted Humphrey. Indianapolis: Hackett.

Sellars, Wilfrid (1962): "Philosophy and the scientific image of man". In: R. Colodny (ed.): *Frontiers of science and philosophy*. Pittsburgh: University of Pittsburgh Press, pp. 35–78.

Contents

1 **Matter in Motion: The Scientific Image of the World**   1
   1.1   Atomism from Democritus to Feynman   1
   1.2   Primitive Ontology   5
   1.3   Dynamical Structure   9
   1.4   Probabilities and the Direction of Time   22
   1.5   Beyond Classical Mechanics: Classical Field Theory   29
   1.6   From Field Theory to Relativity Physics   36
   1.7   From Statistical Mechanics to Quantum Mechanics   44

2 **How Science Explains: Scientific Explanations and Their
   Limits**   63
   2.1   The Location Problem and Its Solution: Functionalism   63
   2.2   What Scientific Explanations Achieve and What Their
         Limits Are   75
   2.3   What Are Laws of Nature?   86
   2.4   Why Determinism in Science Is Not Opposed to Free
         Will   92

3 **Why the Mind Matters: The Manifest Image of the World**   111
   3.1   Sensory Qualities as Problem for the Scientific Image 111
   3.2   Normativity as the Focal Point   120

3.3   The Scientific and the Manifest Image   130
3.4   The Synoptic View   140
3.5   A Twofold Conception of Freedom   157

Summary   163

References   171

Index   183

List of Figures

Fig. 1.1   Configuration of point particles individuated by distance relations   11

Fig. 1.2   Sequence of changing distance relations among a fixed number of permanent point particles with an objective order τ of that sequence. This figure is, however, misleading in that it represents the change as discrete instead of continuous   13

1

Matter in Motion: The Scientific Image of the World

1.1 Atomism from Democritus to Feynman

Science in the Western culture goes back to Ancient Greece, namely the Presocratic natural philosophers. Among them are Leucippus and Democritus (about 400 B.C.), who were the first atomists. Democritus is reported as maintaining that

> ... substances infinite in number and indestructible, and moreover without action or affection, travel scattered about in the void. When they encounter each other, collide, or become entangled, collections of them appear as water or fire, plant or man. (Fragment Diels-Kranz 68 A57; quoted from Graham 2010, p. 537)

In a similar vein, Isaac Newton writes at the end of the *Opticks*:

> ... it seems probable to me, that God in the Beginning form'd Matter in solid, massy, hard, impenetrable, moveable Particles ... the Changes of corporeal Things are to be placed only in the various Separations and new Associations and motions of these permanent Particles. (Quoted from Newton 1952, question 31, p. 400)

© The Author(s) 2020
M. Esfeld, *Science and Human Freedom*,
https://doi.org/10.1007/978-3-030-37771-7_1

To turn to contemporary physics, Richard Feynman says at the beginning of the famous *Feynman lectures*:

> If, in some cataclysm, all of scientific knowledge were to be destroyed, and only one sentence passed on to the next generations of creatures, what statement would contain the most information in the fewest words? I believe it is the *atomic hypothesis* (or the atomic *fact*, or whatever you wish to call it) that *all things are made of atoms—little particles that move around in perpetual motion, attracting each other when they are a little distance apart, but repelling upon being squeezed into one another*. In that one sentence, you will see, there is an enormous amount of information about the world, if just a little imagination and thinking are applied. (Feynman et al. 1963, ch. 1–2)

This is atomism. The success story of modern science is at its roots the success story of atomism. It is evident from these quotations why atomism is attractive: on the one hand, it is a proposal for a theory about what there is in the universe that is both most parsimonious and most general. On the other hand, it offers a clear and simple explanation of the realm of the objects that are accessible to us in perception. Any such object is composed of a large number of discrete, pointlike particles. All the differences between these objects—at a time as well as in time—are accounted for in terms of the spatial configuration of these particles and its change. This view is implemented in classical mechanics. It conquered the whole of physics via classical statistical mechanics (e.g. heat as molecular motion), chemistry via the periodic table of elements, biology via molecular biology (e.g. molecular composition of the DNA), and finally neuroscience—neurons are composed of particles, and neuroscience is applied physics. In a nutshell, what paved the way for the success of science is the idea to decompose everything into elementary particles and to explain it on the basis of the interactions of these particles.

To understand how the atoms interact, one needs laws that describe their motion. That is why atomism remains a speculative stance in Antiquity and becomes science only in modern times: only modern physics formulates laws of motion for the atoms. Nonetheless, the attractiveness of atomism does not depend on what precisely is proposed as these laws. Its attractiveness is independent of a particular physical theory. It consists in the idea of composition by particles together with the idea that differences in this composition account for all the differences that there are. There is a direct and intuitive link from this idea to the observable, macroscopic objects.

That link is direct and intuitive because all that is observed in science as well as in common sense are the positions of discrete objects relative to each other and the change of these positions—in other words, the variation in the distances among discrete objects that make up a configuration of objects and the change of such configurations. Accordingly, all measurement outcomes are recorded as relative positions within configurations of discrete objects and variations of such positions, such as, for instance, pointer positions or digital numbers on a screen. In this vein John Bell (2004, p. 166) famously says "… in physics the only observations we must consider are position observations, if only the positions of instrument pointers". The qualification "in physics" is appropriate, because common sense observations typically involve colours, sounds or scents of spatially arranged objects. The positions of objects are discerned by means of these sensory qualities. However, sensory qualities do not figure in physical theories, at least not explicitly (we will consider that issue in Sect. 3.1).

That notwithstanding, all the evidence that we have in science is evidence of positions of discrete objects relative to other discrete objects. For instance, even in the case of the gravitational waves detected by LIGO (Laser Interferometer Gravitational-Wave Observatory) in 2016, all the evidence is evidence of change in the relative positions of discrete objects that finally are particles. This change then is mathematically described in terms of a wave rippling through the gravitational field. This fact highlights again the direct link between the experimental evidence and the idea of atomism: it is relative positions of discrete objects all the way down from the macroscopic objects to their ultimate constituents, or all the way up from the ultimate constituents to the macroscopic objects. Thus, if a theory gets the spatio-temporal arrangement of the particles right (that is, the arrangement of fermionic matter according to contemporary physics),[1] it has got everything right that can ever be

---

[1] Cf. Bell (2004, p. 175).

checked in scientific experiments.[2] Two theories that agree on the spatio-temporal arrangement of the particles cannot be distinguished by any empirical means, whatever else they may otherwise say and disagree on. By the same token, two possible worlds with the same spatio-temporal arrangement of the particles are indiscernible by any scientific means.

Hence, what is relevant for the account of the perceptible macroscopic objects and their differences are only the relative positions of the particles—in other words, how far apart they are from each other, that is, their distances—and the change of these distances. Any intrinsic nature of the atoms is irrelevant for that task. Realizing this point stands in contrast to the mainstream tradition in ancient and medieval thought where the focus was on an inner form (*eidos*) of the objects—that is, some characteristic, intrinsic features that belong to each object considered independently of all the other objects. Aristotle's *Categories* and *Metaphysics* are the *locus classicus* of this tradition. To put it differently, on atomism, the atoms are the substance of the world. They are permanent: they do not come into existence and they do not go out of existence. But they are substances only in the sense of permanent existence. They are not substances in the sense of having an inner form. The atoms are featureless. All there is to them are their positions relative to each other—that is, their distances—and the change of these positions.

René Descartes is the central figure who brought about the shift from Aristotelian forms in the medieval, scholastic conception of nature to an essence of the material objects that consists only in their extension—that is, the spatial relations or distances among these objects—and motion (that is, the change of these spatial relations). In short, for Descartes, nature is only *res extensa*. Descartes also formulated laws of motion. But these did by and large not turn out to be correct, mainly because Descartes conceived the interaction of the material objects in a mechanical way as direct contact. Laws of interaction that prevailed go back to Newton, with the law of gravitation being the prime example. Newtonian gravitation is interaction without direct contact, as in the attraction of the Earth by the Sun. Let us therefore have a closer look at the interplay between objects and laws.

---

[2] See also Maudlin (2019, pp. 49–50).

1.2   Primitive Ontology

The atoms that atomism poses cannot be further decomposed into smaller things, because they are not extended themselves: they are point particles. All the extension comes from the spatial relations in which they stand, making up for configurations of point particles. These are the bedrock of the universe so to speak, since one cannot go further down than spatially arranged point particles in scientific enquiry. In other words, their configurations are the ultimate referents of our scientific theories, what they talk about in the last resort. Let us introduce the philosophical term *primitive ontology*. Ontology is about what there is (*to on* in ancient Greek). The primitive ontology is about what is admitted as simply existing in the sense that it cannot be derived from anything else or introduced in terms of its function for anything else. What takes this place depends on our theories: it is the hypothesis of science that the universe is ultimately constituted by spatially arranged point particles. If this hypothesis is right, then the particle configuration of the universe is the bedrock, at least as far as scientific enquiry is concerned.

Are there alternatives to atomism? The Presocratic natural philosophers do not only include the atomists Leucippus and Democritus. Before them came Thales, Anaximander, Anaximenes and Anaxagoras who searched for the stuff out of which everything is made. Thales apparently took water to be that stuff, whereas the others thought of it as something more abstract. In any case, the stuff view of nature is opposed to atomism: instead of a plurality of discrete, indivisible objects, there is just one continuous stuff that stretches out throughout the universe. One problem with this view is that one may find the idea of a bare stuff substratum of matter mysterious. Furthermore, that stuff substratum admits of different degrees of density as a primitive matter of fact: there is more stuff in some regions of space than in others. In brief, there is nothing in this view that individuates or distinguishes material objects by properties or relations, such as their spatial relations in a configuration of discrete objects as on atomism.

More importantly, it is unclear how this view could account for the macroscopic world with which we are familiar. There is nothing in this view that matches the theory of composition in atomism: the spatially arranged point particles compose what is known today as atoms in the sense of the chemical elements, these compose molecules, and the molecules finally compose the macroscopic objects with which we are familiar. In a nutshell, water is not a continuous, primitive stuff, as the ancient conception of the four elements earth, water, air and fire has it. Water consists in molecules that are composed of hydrogen and oxygen atoms, which, in turn, are composed of protons, neutrons and electrons, etc. until one gets down to the point particles.

The primitive ontology hence is not a matter of speculation. Although whatever is supposed to be the bedrock of the universe is likely to be quite far away from the features of the world with which we are familiar, there has to be a clear and intelligible link with these features, such as the link from particles to macroscopic objects via composition. Nonetheless, the crucial point is not the idea of composition as such, however intelligible or intuitive it may be, but to cash out the promise of explaining all the differences in the macroscopic objects in terms of differences in the particle composition and change in that composition. Laws of motion for the particles are indispensable to achieve this aim. That is why this aim is achieved only by modern science. Consequently, both atomism and the view of a continuous stuff remain speculative before the advent of modern science. Modern science then vindicates atomism by providing laws of particle motion on the basis of which the promise of explaining the differences in the macroscopic objects in terms of differences in the particle configuration can be fulfilled.

Stressing the importance of laws brings out that the intuitive link from spatially arranged point particles to macroscopic objects via composition is not the argument for the primitive ontology of atomism. The fact that all that is observed in science and common sense are spatial arrangements of discrete objects and the change of these arrangements is not the argument for that ontology either. This fact and that intuition suggest trying out a primitive ontology of point particles that are characterized by their relative positions and the change of these positions only. But the argument for that ontology then is its explanatory force, that is, how it accounts for all the evidence that we have on that parsimonious basis. In order to spell out such an account, formulating laws of particle interaction and testing them is crucial.

Even if all that is pertinent for scientific explanations are the relative positions of the point particles only, one may wonder whether there has to be more to them than relative positions for them to be the substance of the world—that is, the objects that compose everything else. In other words, even if an intrinsic essence of the atoms is irrelevant for and inaccessible to science, it may nevertheless have to exist for the atoms to be able to do what science wants them to do. And if there is not an intrinsic essence of the individual atoms, it seems that there has to be at least a general stuff essence of matter—something more than relative positions in virtue of which the atoms are material objects. Otherwise, it seems that their material nature fades away upon inquiry, if all that remains of matter is the geometry of distances between sparsely distributed point particles as well as the change of these distances. However, this concern is misplaced.

If there is a plurality of objects, there has to be something that individuates them—that is, something that answers the question why this is one object, that another object, etc. so that there really is a plurality of objects instead of just one object. Furthermore, there also has to be something that unites these objects so that they make up a world. In other words, there has to be a world-making relation, that is, a relation that binds all and only those objects together that make up a world. It is evident that the distance relation performs the latter task: all and only those objects that are spatially related constitute a world. If there were objects that are not at a distance to each other, they would inhabit different worlds. If they are related by a distance, they are in one and the same world, as stressed by David Lewis (1986a, ch. 1.6) among others.

Moreover, the distance relations—and only they—individuate the objects: what distinguishes each object in a configuration of objects is the position that it has relative to all the other objects. Even if a configuration is partially symmetrical, there always is at least one object outside that symmetry relative to which all the other objects can be distinguished. Thus, for example, motion can always be referred to the fixed stars as reference system relative to which the other objects are in motion and can be distinguished by their distance. Entirely symmetrical configurations may be imaginable (provided that one admits tacitly the intuition of an absolute space in which they are embedded). However, according to Leibniz' principle of the identity of indiscernibles, they do not represent a possible world, because there would be nothing in such configurations that individuates their elements. Consequently, it would be an illusion to think that there is a plurality of elements or objects—and hence a configuration of something—in such a case.[3]

Scientific parameters that are attributed to physical objects over and above their relative positions—such as mass or charge—cannot distinguish them: they differentiate between various kinds of particles, such as the particle species admitted in today's standard model of elementary particles. But they cannot distinguish between the individual particles within a species or kind, because all the particles of a given species—such as, for instance, all electrons—have the same values of mass, charge, etc. Consequently, there are no qualitative properties that could individuate the particles. Hence, the demand for something that individuates the physical objects is fulfilled by the distance relations, and only by them. Therefore, there is no need for anything more than distance relations to both individuate the objects and have a relation that binds them together so that they constitute a world. This insight is expressed in today's metaphysics by the stance that is known as *ontic structural realism*.[4]

However, is this enough to characterize the particles as *material* objects? As already mentioned, Descartes defined matter as *res extensa*. Indeed, that is what matter is according to science. There is nothing more to matter beyond extension in the guise of distance relations between point particles and their change. In particular, there is no stuff-essence of matter. It is entirely unclear what such a stuff-essence could be. The impenetrability of matter, often invoked as criterion that characterizes matter, is accounted for by the individuation of the material objects through the distance relations as well: for there to be two material objects, there has to be a distance between them—that is, a non-vanishing distance; consequently, if there are two objects, they cannot penetrate each other.

---

[3] Cf. Hacking (1975) and Belot (2001).
[4] See notably Ladyman (1998), French and Ladyman (2003), Esfeld (2014b), Esfeld and Lam (2008) and French (2014).

Furthermore, the definition of matter in terms of distance relations is substantial: it distinguishes material from non-material objects. Descartes defined matter as *res extensa* and mind as *res cogitans*.[5] That is to say: standing in distance relations (extension) makes it that points are matter points (point particles), whereas standing in thinking relations makes it that points are minds. This is not to say that there are indeed fundamental thinking relations and fundamental mind points. Everything may be material. But whether or not this is so is not a matter of definition. It has to be established by argument. We will go into this issue in Chap. 3. What is important for now is that distance relations provide for a nontrivial and operational definition of matter that distinguishes matter from entities that are not matter (independently of whether or not such entities exist).

To sum up, we can formulate the primitive ontology proposed by atomism in terms of the following two axioms or principles:

1. *There are distance relations that individuate simple objects, namely point particles (matter points).*
2. *The point particles are permanent, with the distances between them changing.*

Esfeld and Deckert (2017, ch. 2) is a detailed discussion of these axioms.

1.3 Dynamical Structure

Can there be only distance relations—that is, spatial relations—without there being a space in which these relations are embedded? And can there be change in these relations without a time in which that change takes place? Isaac Newton is well-known for posing absolute space and absolute time in his masterpiece *Mathematical principles of natural philosophy* (1687; see the scholium to the definitions). Thus, space and time exist independently of matter. They are like a container in which the material objects are inserted and in which the history of the universe plays itself

---

[5] See Descartes, *Principles of philosophy*, in particular part 1, § 53.

out. Gottfried Wilhelm Leibniz famously attacked Newton on this issue, maintaining, expressed in today's terms, that absolute space and time are a *surplus structure*: they allow for much more possibilities than those that can be distinguished empirically. For instance, if the whole configuration of matter of the universe is embedded in Newtonian absolute space, it can be shifted, or rotated as a whole, in that space. However, no such shift or rotation would make any empirical difference: all the relations between the material objects would remain the same. It would be a difference (different positions of all the material objects in absolute space) that does not make a difference.[6]

According to Leibniz, space is the order of what coexists—that is, particles bound together by distance relations to make up the world. Hence, there are only spatial relations. Space does not exist. It is only a means to represent how these relations bind objects together so that these objects coexist. In other words, the objects are point particles and hence not extended themselves. Extension exclusively consists in the distance relations between them. These relations are not composed of anything, in particular not of points. Since there is no space, there are no points of space either. Time is the order of succession: the distance relations among the material objects change. Time is the means to represent that change by imposing a metric on it. Hence, change is basic, and time is derived from it.[7]

The distance relation has several characteristic features. In the first place, it is irreflexive: nothing can stand in a distance to itself. It is symmetric: if matter point i is at a certain distance to matter point j, then j is at the same distance to i. It is connex, meaning that any two objects in a configuration stand in a distance relation to each other. It fulfils the triangle inequality—that is, for any three matter points i, j, k, the sum of the distances between i and j and j and k is greater than or equal to the distance between i and k. What is important are the ratios between the distances—that is, not how far is i from j in absolute terms, but how far is i from j in comparison to how far is i from k, and k from j. To carry out

---

[6] See Leibniz, third letter to Newton-Clarke, § 5, in Leibniz (1890, pp. 363–364); English translation Leibniz (2000).

[7] See Leibniz, third letter to Newton-Clarke, § 4, fourth letter, § 41, fifth letter, §§ 29, 47, 104, in Leibniz (1890, pp. 363, 376–377, 395–396, 400–402, 415).

such comparisons, one needs numbers, more precisely real numbers, as means of representation. However, employing numbers does not smuggle in any infinity into the ontology. Numbers are only a means of representation; they do not belong to the ontology. The ontology of the universe are very many matter points. If there are N matter points (at a minimum 3), there are finitely many distance relations, namely $1/2N(N-1)$ distance relations.

Obviously, there is more to the concrete distance relations that exist in the universe than just fulfilling the triangle equality.[8] Nonetheless, all there is to them can be expressed in terms of scale invariant quantities. Again, only the ratios between the distances enter into the ontology; scales are a means of representation. Thus, the most elaborate proposal how to understand physics on the ontological basis of distance relations among objects only is the relational mechanics or shape dynamics developed by Julian Barbour and collaborators since the 1970s.[9] All that there is to the particles is contained in the shape of the particle configuration, which requires endorsing distances and angles that define the shape as primitive, but which is scale invariant. Hence, it makes no sense to ask how far apart a particle is from other particles in an absolute sense, but only what the ratio of the distances among the particles is (Fig. 1.1).

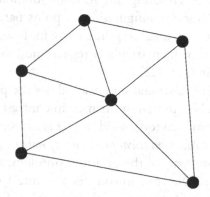

Fig. 1.1 Configuration of point particles individuated by distance relations

---

[8] Lazarovici (2018b) is right in laying stress on this issue.
[9] See Barbour (2012) and Mercati (2018) for an overview.

As regards time, Leibniz's view of time as the order of succession is what is known as a causal theory of time: there is change as a primitive matter of fact. On that basis, the time parameter enters as means to measure the change. The term "causal theory of time", however, is misleading in that there is no question of causes of change at this stage. The point at issue here only is that change in the spatial configuration of matter is primitive; there is no external time in which it takes place. Nonetheless, change is directed: it goes from one particular arrangement of the distances among the particles to another such arrangement within the particle configuration of the universe. Consequently, there is an objective history of the universe, namely an objective sequence in which the distances between the point particles in the universe evolve.

However, it makes no sense to ask how long such an evolution takes or how fast it goes. The presupposition that the dynamics of the universe develops according to an external time has no physical meaning. If the entire universe could evolve at different external time rates, then any two such evolutions would be physically indistinguishable: they would consist in exactly the same sequence of changing distance relations among the point particles. That is why absolute time, like absolute space, implies a commitment to a surplus structure. One can employ time as the measure of change only by choosing one subconfiguration of distance relations within the universal configuration of point particles as the clock relative to which one then measures the change in the other relations. The clock is distinguished by its particularly regular motion in comparison to the measured motion (Fig. 1.2).

In brief, Leibnizian relationalism encapsulates the primitive ontology of atomism. Also in Newtonian mechanics, in contrast to the impression that one may get from Newton's words if one reads only the scholium to the definitions in the *Mathematical principles of natural philosophy* (1687), space and time are not part of the primitive ontology. As is evident from the context in which Newton introduces absolute space and time, he relies on these notions for reasons of the dynamics of matter, namely to formulate his famous laws of motion.

In order to fulfil the promise of atomism to account for all the differences among bodies in terms of particle configurations and their change, Newton conceives a reference state for the particles relative to which all

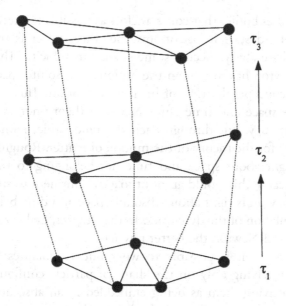

Fig. 1.2 Sequence of changing distance relations among a fixed number of permanent point particles with an objective order τ of that sequence. This figure is, however, misleading in that it represents the change as discrete instead of continuous

the other states are then conceptualized. This reference state is formulated in the first law of motion:

> Every body continues in its state of rest, or of uniform motion in a right line, unless it is compelled to change that state by forces impressed upon it. (Quoted from Newton 1934, p. 13)

This is the state of inertial motion, that is, motion without change in velocity or direction. But the question is: rest or uniform motion with respect to what? What defines the rest, the right line and the constant velocity? This is where absolute space comes in: the first law conceptualizes rest and inertial motion as motion with respect to absolute space. Newton needs absolute space in order to define a reference state for the motion matter.

Obviously, motion with respect to absolute space cannot be measured, and there is no completely uniform motion in the universe. When doing physics, one uses bodies that come as close as possible to inertial motion, such as the fixed stars in astronomy. But there never is truly inertial motion, as there always occurs some change in velocity. That notwithstanding, Newton has to rely on the notions of absolute space and time for the very conceptualization of his laws of motion. This fact confirms that absolute space and time enter Newton's theory not as part of the primitive ontology, but through their dynamical role, namely the role that they play for the theory of the motion of matter. Roughly speaking, Leibniz is right about space and time not belonging to the primitive ontology (because this would amount to a commitment to surplus structure), and Newton is right about absolute space and time being required for the formulation of the dynamics. In that respect, Leibniz is the better philosopher and Newton the better physicist.

More precisely, it is possible to work out a dynamics for classical mechanics by relying only on the shape of particle configurations and without conceiving them as being embedded in an absolute space and time, that is, without employing absolute, scale-dependent quantities in the formulation of the dynamics, such as absolute acceleration. Barbour and his collaborators did so in detail.[10] Hence, the dynamics does not have to rely on quantities that refer to an absolute space and time. The possibility of such a dynamics corroborates the stance of a primitive ontology that admits only distance relations and their change. However, by refusing to employ absolute spatio-temporal quantities, the result is a mathematical formulation of the dynamics that is more complicated than the Newtonian one. In sum, absolute space and time and related notions are useful for the formulation of a dynamics, because they simplify the dynamics; but they lead to trouble if one includes them in the ontology, that is, if one admits them to what one takes to exist in the world.

Given the reference state in the guise of inertial motion, Newton then accounts for every deviation from inertial motion—and thus every change in the motion of an object—in terms of forces acting on that object. The operation of forces is stated in Newton's second law of motion:

---

[10] See Barbour and Bertotti (1982), Barbour (2003) and Mercati (2018, part II).

The change of motion is proportional to the motive force impressed; and is made in the direction of the right line in which that force is impressed. (Quoted from Newton 1934, p. 13)

The third law then adds that to every action of a force on an object corresponds a reaction from that object:

To every action there is always opposed an equal reaction: or, the mutual actions of two bodies upon each other are always equal, and directed to contrary parts. (Quoted from Newton 1934, p. 13)

These three laws are a framework in which precise laws of motion can then be formulated. Unless we are told which are the forces that operate in nature, we cannot formulate a law that allows to calculate the motion of bodies. The only force that Newton treats in that respect is gravitation. His law of gravitation is a breakthrough because it unites astronomy and mechanics: the law applies on Earth—for instance to the apples falling from the tree in front of Newton's office in autumn—as well as in the Heavens, for instance to the motion of the Earth around the Sun. To calculate the action of a force, that is, the change in the motion of an object brought about by the force, one has to know the position and the velocity of the object as initial condition. Moreover, the force is such that it relies on at least one further parameter attributed to the objects. In the case of gravitation, this is mass.

Over and above position and velocity, the particles in Newtonian mechanics have a mass, which is differentiated into inertial mass and gravitational mass, whose values are the same and always remain constant for any given particle. However, this does not make mass an intrinsic property of the particles, something that a particle has in and of itself, independently of all the other particles. Mass is a dynamical parameter that is defined in terms of its function for the particle motion, namely the resistance to acceleration due to inertial mass and attractive motion due to gravitational mass. As Ernst Mach says when commenting on Newton's *Principles*, "The true definition of mass can be deduced only from the dynamical relations of bodies" (1919, p. 241). Hence, mass is a parameter that expresses a dynamical relation among the physical objects.

Once this parameter is given, there is no further force parameter in the law of gravitation. The gravitational attraction between any two bodies at any given time is fixed by the relative positions of these bodies—that is, their distance, with the attraction going down the greater the distance is—, their masses and the gravitational constant as well as their velocities as initial condition. In general terms, the change in velocity of the objects in the universe at any given time is fixed by their positions, velocities and masses modulo the gravitational constant. If one knows these values, one knows the change in velocity of the objects. The same applies to all the other forces: they derive from a specific dynamical parameter attributed to the objects, plus their positions and velocities as initial conditions (modulo further constants of nature). In the case of the gravitational interaction, the specific dynamical parameter is gravitational mass; in the case of the electromagnetic interaction, the specific dynamical parameter is the charge of the particles. That is to say: "force" in Newtonian mechanics is a placeholder for whatever parameter is attributed to the particles in terms of its precise function for the change of the motion of the particles (such as mass, or charge).

This fact shows that it is wrong-headed to conceive forces as some sort of physical things in nature that literally act on the particles by accelerating them. Such a view is a residue of the mechanical conception of interaction through direct contact: if it is a fact that there is interaction without direct contact between the objects, it is then supposed that there is a force that goes from one object to other objects when they interact. Science, however, teaches us that interaction is much more abstract than any form of direct contact: *interaction is correlated change of the motion of objects*, such as two objects moving towards one another as in gravitational attraction. This correlated change is traced back to parameters that are attributed to the objects (such as mass and charge) and certain constants (such as the gravitational constant), with these parameters and constants being introduced in terms of their functional role for the change in the motion of the objects. But there is nothing that brings about this correlation by being transmitted between the objects. In Newtonian physics, this fact is usually described in terms of action at a distance: given the positions, velocities and masses of two bodies at any time, their acceleration *at that very time* is fixed. Hence, there is no time for something like a force to be transmitted from one object to the other one. However, the term "action" is misleading, when it is taken to suggest that something acts instantaneously across space, instead of there being just correlated motion of objects.[11]

All the parameters that a physical theory introduces over and above those ones that define the primitive ontology can be considered as constituting the *dynamical structure* of the theory. The primitive ontology can be identified in terms of those parameters that are not introduced through their function for something else, thus referring to what there simply exists in the world according to the theory. The position of objects is the prime—and arguably the only—example of such a parameter. Thus, in atomism, the primitive ontology is point particles individuated by their relative positions in a configuration (that is, the distances among the point particles) and change of these positions (that is, motion of the point particles). However, a configuration of point particles defined in terms of their relative positions does not contain information about its evolution: given any such configuration (as illustrated in Fig. 1.1 above), one cannot infer how it will develop. To put it in a nutshell, relative positions at a time do not contain information about motion. In other words, if one considers the primitive ontology as defining the kinematics of a theory, then the kinematics does not enable inferences about the dynamics (apart from the fact that the number of particles and thus their individuation through the distance relations has to be preserved: the evolution to an entirely symmetrical configuration is thereby excluded in Leibnizian relationalism).

Given the fact of change the task of science is to discover patterns in the sense of regularities in the change of the relative positions of the particles. Only such regularities then enable the explanation of the differences in the perceptible macroscopic objects in terms of differences in the arrangement of the microscopic particles that compose them and thus to fulfil the promise of atomism. These regularities are the particle interactions in the sense of correlated motions. To represent correlated motions, more parameters are needed than relative positions and change of positions (which can be represented in terms of velocity, that is, the

---

[11] For an excellent discussion of these issues, see Lange (2002, ch. 1).

first temporal derivative of position). The question is this one: What fixes the velocity, including the change in velocity?

Everything that enters into the answer that a physical theory gives to this question is part of the dynamical structure of the theory. If there are only relative distances and not an absolute space, patterns or regularities in particle motion can only be correlations in the particle motions, that is, particle interactions. Accordingly, the meaning of the term "structure" in "dynamical structure" are correlations that are defined in terms of dynamical parameters on the objects (i.e. the particles) such that these correlations constitute a restriction of the possible motion of the particles. For instance, roughly speaking, if the particles have gravitational mass, only certain attractive motions are possible; if they have charge, only certain attractive and repulsive motions are possible, etc.

The task hence is to conceive parameters that can be introduced in terms of their functional role for fixing the evolution of the particle positions (the primitive ontology). This functional role is also known as causal role, although, again, this term can be misleading, since no specific notion of causation is involved here, in particular not a notion of bringing about or producing something. The point at issue is the conceptualization of patterns of motion in the sense of correlated motions. These parameters then enable the formulation of a law for that change, such as the law of gravitation in terms of particle mass and gravitational constant.

Patterns of motion can be conceptualized in the guise of a first order mathematical theory with parameters that directly fix the velocity of the particles given their positions as initial condition (first order because velocity is the first temporal derivative of position). An example is the wave-function on configuration space in quantum mechanics. It can also be done in the guise of a second order mathematical theory that demands over and above position a velocity as initial condition and is then concerned with the change of velocity as a function of additional parameters (acceleration as the second temporal derivative of position). The paradigmatic example is Newtonian mechanics with inertial motion (constant velocity) as the reference state and forces being the placeholder for the parameters that are introduced in terms of their function for the change in the velocity of the particles, such as gravitational mass.

A theory can introduce whatever dynamical parameters serve that purpose, that is, constant parameters—such as mass, charge, total energy, constants of nature—or parameters at an initial time (such as momenta, fields, an initial wave-function). The benchmark for these parameters is to lead to the formulation of laws of motion such that given an arbitrary configuration of matter as input, the law yields the evolution of that configuration as output. The geometry also serves that purpose. Thus, the geometry of an absolute, Euclidean space in which the configuration of matter of the universe is represented as being embedded and an absolute time in which it is represented as evolving are an essential means for the formulation of the Newtonian laws of motion. Indeed, the geometry, the dynamical parameters and the laws come as a package: the precise functional definition of the dynamical parameters involves the law, and the law is formulated by using the dynamical parameters as well as the geometry. But there is no threatening circularity here: roughly speaking, all three are conjectured at once and then made precise together in order to achieve a theory that tells us how the particles move.

In fundamental physics—that is, in physical theories that can no longer be traced back to other scientific theories—, the dynamical structure concerns automatically the universe as a whole. Strictly speaking, the laws of such theories relate only the state of the whole universe at one time with the state of the whole universe at other times. In other words, they relate one arrangement of the matter of the universe with all the other arrangements of the matter of the universe, which are generated by the change in the configuration of matter of the universe. Consider again Newtonian mechanics: although the gravitational attraction is calculated in each case only for pairs of objects, strictly speaking, the change in the velocity of any one object in the universe at any one time depends on the positions, velocities and masses of *all* the other objects in the universe at that time via the law of gravitation, since all the objects interact with each other.

Also this fact is well brought out in a famous statement by Pierre Simon de Laplace:

We may regard the present state of the universe as the effect of its past and the cause of its future. An intellect which at a certain moment would know all forces that set nature in motion, and all positions of all items of which nature is composed, if this intellect were also vast enough to submit these data to analysis, it would embrace in a single formula the movements of the greatest bodies of the universe and those of the tiniest atom; for such an intellect nothing would be uncertain and the future just like the past would be present before its eyes.[12]

Thus, the dynamical structure of a fundamental physical theory relates the state of the universe at one time to the state of the universe at other times. However, the intellect that Laplace imagines here cannot be a being within the universe. No being within the universe could know the initial conditions of all the objects in the universe including itself.[13]

We hence face the following paradox: on the one hand, the dynamical structure of a fundamental physical theory is defined for the universe as a whole. On the other hand, it is in principle not possible for a being within the universe to know the initial conditions of the universe as a whole. Consequently, no being within the universe could ever solve the dynamical equations of a physical theory insofar as these are defined for the universe as a whole. The way out of this paradox consists in three steps: (1) The very dynamical structure of a fundamental physical theory that is defined for the universe as a whole has also to be applicable to particular subsystems within the universe, notably subsystems that are within our cognitive reach. (2) In order for the application of the dynamical structure to subsystems to be successful, one has to presuppose that certain stable conditions obtain in the environment, whereby the environment is the rest of the universe outside the subsystem under consideration. (3) One assumes that the dynamics of the subsystem under consideration is insensitive to slight variations in its initial conditions. Only if these three conditions are satisfied does Laplace's ideal of a deterministic law—that is, a law fixing the whole evolution of the system given initial conditions—enable deterministic predictions.

The first condition can be assured simply by theory construction: in setting up the dynamical structure of a physical theory, one can make

---

[12] Quoted from Laplace (1951, p. 4); original publication 1814.
[13] See e.g. Popper (1950a, b) and Breuer (1995).

sure that this very structure is applicable not only to the universe as a whole, but also to subsystems within the universe, or at least contains a procedure how to derive its application to subsystems. Consider again Newtonian gravitation: the theory says that the gravitational acceleration of any one object in the universe at any given time depends on the positions, velocities and masses of all the other objects in the universe at that time. But it formulates that dependence in mathematical terms of correlated motion of pairs of objects.

However, in applying the law of gravitation to pairs of objects, one has to presuppose that nothing outside the pair in question interferes in its interaction in a significant manner. In other words, one has to presuppose that the influence of the environment (which is the rest of the universe) can be neglected, at least for all practical purposes. The satisfaction of this condition cannot be assured by theory construction. It is a substantial assumption about what the world is like. In Newtonian gravitation applied to our universe, fortunately, for example, there usually do not come in heavy objects from far apart in space that interfere in a considerable manner with the trajectory of the Earth around the Sun as determined only by the parameters that apply to Earth and Sun, or with the trajectory of a stone that falls on the Earth as determined only by parameters that apply to the stone and the Earth. Fortunately furthermore, these trajectories are by and large insensitive to slight variations in the initial positions and velocities of these pairs of objects. That is why in these paradigmatic cases, one can apply the Newtonian law of gravitation to make deterministic predictions about the motion of particular objects, although one does not know the rest of the universe and one is in fact not in the position to know their initial positions and velocities exactly.

Whereas the second condition—no significant influence from objects that are far apart in space—is usually fulfilled in the universe, the third condition is in fact satisfied only in rather rare cases. In many, if not most cases, slight variations in the initial conditions have huge consequences for the future evolution of the objects under consideration. In these cases, then, deterministic laws do not enable deterministic predictions. That is why over and above a primitive ontology and a dynamical structure we need probabilities as the third element in order to understand the success of modern science.

1.4   Probabilities and the Direction of Time

Since Antiquity, the regularities in the motion of the planets are well known. What was in general not realized in Antiquity is that the same regularities apply also on Earth, such as when throwing a stone to the soil, or when an apple falls from a tree. Such regularities are the evidence on the basis of which a universal law such as the law of gravitation is conceived: they are paradigm examples of applying the law by making deterministic predictions of single cases that are confirmed by observations. However, there also is evidence of regularities that do not apply to single cases, but that show up only when one considers large ensembles of similar events. Consider instead of a stone thrown to the soil a coin thrown to the soil. In the case of the coin throw by contrast to the stone throw, the outcome is highly sensitive to slight variations in the initial conditions: a minimal variation in the way in which the coin is thrown (i.e. a minimal variation in its initial position or initial velocity) can change the entire outcome, that is, whether the coin lands heads or tails. The prediction of the result of an individual coin toss therefore is practically impossible. Nonetheless, the coin toss comes under the law of gravitation in the same way as does throwing a stone, or planetary motion. The difference between these cases lies only in the sensitivity to slight variations in the initial conditions.

That sensitivity notwithstanding, there also is an observable regularity in the coin toss case, namely a statistical one: if one considers a large series of coin tosses under the same circumstances, such a series typically tends to manifest an equal distribution of heads and tails. This then is the basis for the statistical prediction that the probability for the outcome "heads" and the probability for the outcome "tails" is 0,5 each (i.e. 50% chance heads, 50% chance tails). One cannot obtain these probabilities from classical mechanics alone. Over and above the laws of classical mechanics, one needs a probability measure: a large series of coin tosses that fails to converge towards an equal distribution of heads and tails neither contradicts the laws of classical mechanics nor those of probability theory. Such a series is not impossible given the laws. It is atypical. Saying that a series of coin tosses typically converges towards an equal distribution of heads and tails the larger it gets means that for the by far vast majority of possible initial conditions of coin tosses, this distribution of the outcomes obtains. To take another example, consider a gas contained in a box with a barrier in the middle of the box such that the gas is concentrated in one half of the box. If one removes the barrier, the gas will typically expand throughout the whole box, evolving towards an equilibrium state with an even distribution of the molecules throughout the box. That is to say, for the vast majority of possible initial conditions, such an evolution obtains.

In order to make such statements, one has to have the means to represent the possible initial conditions of the objects under consideration in a mathematical space and then to define a typicality measure on that space, that is, a measure that makes a statement about what holds for the vast majority of initial conditions for specific systems in the universe. This was done in the nineteenth century when Newtonian mechanics was cast in a formulation introduced by William Rowan Hamilton that conceives what is known as phase space to represent the particle configuration and its evolution; for N particles, phase space has $6N$ dimensions—three ones for the initial position and three ones for the initial velocity of each particle. Thus, each point of $6N$-dimensional phase space represents a possible configuration of N particles in three-dimensional physical space. The Lebesgue measure is used as typicality or probability measure on phase space. This then enables the formulation of statements such that gases typically evolve towards an equilibrium state of even distribution in a container, etc., which make predictions about the statistical distribution of outcomes possible.[14]

The probability or typicality measure is not contained in the laws. But it has to be closely related to them: the laws are there to capture observable regularities that often are statistical regularities. It makes no sense to have a simple and elegant theory linked with a typicality measure such that what happens actually in the universe comes out as atypical in the light of that measure. Such a theory would not make what actually happens in the universe intelligible, since it would not make the observed statistical regularities intelligible. Hence, laws and initial conditions are

---

[14] See Lazarovici and Reichert (2015) for a detailed presentation.

linked up with one another. If there is no path from the laws to deterministic predictions because the evolution of the objects under consideration is highly sensitive to slight variations in the initial conditions, the laws have to allow the definition of a probability or typicality measure that applies to the vast majority of the initial conditions in which the objects under consideration actually are. In this manner, the laws enable statistical predictions and thereby makes the observed statistical regularities intelligible.

This consideration confirms that probabilities are objective and indispensable, even if deterministic laws apply to the universe. Probabilities are objective when there are statistical regularities in the world. These regularities do not depend on the beliefs of persons. For instance, it is a fact about the world that the vast majority of gases evolve towards an equilibrium state. In short, when the laws are deterministic, probabilities come in through our ignorance of the exact initial conditions of the subsystems in the universe under consideration. That notwithstanding probabilities are objective, because they concern statistical regularities that obtain in the world independently of the beliefs of persons.

The physical theory that is concerned with probabilities is statistical mechanics, which was developed in the nineteenth century and which is derived from classical mechanics in its Hamiltonian formulation. In fact, statistical mechanics made a central contribution to atomism becoming the paradigm of science due to its explanation of thermodynamical phenomena such as heat in terms of particle motion. Thermodynamical processes are moreover a paradigm example of macroscopic irreversible processes. Gases typically evolve towards an equilibrium state and remain in that state. To consider a common sense example, if one pours cold milk into a cup of hot coffee, the milk and the coffee will rapidly mix, the temperature will rapidly become the same throughout the whole drink and the drink will remain in that state of equilibrium. There will neither be a hot part of the drink with coffee at the bottom of the cup and a cold part at the top with the milk nor again an evolution to such a state. To take another example, if a glass falls from the table down on the floor, it is and remains broken. The scattered pieces of the glass will not spontaneously come together and form a glass again.

In thermodynamics, these are phenomena that come under the second law, which states an increase in entropy, that is, roughly speaking, an increase in disorder in the sense of uncoordinated motion. In statistical mechanics, these phenomena are explained in terms of molecular motion: the laws of classical mechanics allow in principle for all processes to be reversed. However, the initial conditions in the form of a coordination of the initial positions and initial velocities of the atoms such that the milk molecules and the coffee molecules in the drink remain separated (or separate again), or such that the pieces of the broken glass spontaneously come together to form a glass again, are quasi never satisfied, although they are allowed by the theory. If one considers the possible initial conditions in phase space, the by far vast majority of initial conditions is such that the drink will quickly move into thermal equilibrium, the pieces of the broken glass remain scattered, etc. Given the Lebesgue measure on phase space, these processes come out as typical, and probabilities for the evolution of these objects can be calculated without taking into account the exact initial conditions that obtain in each individual case.

This reasoning was worked out mainly by Ludwig Boltzmann (1896/1998, English translation 1964). It amounted to a vindication of atomism: for the first time, a physical theory concerned with macroscopic phenomena that was not formulated in terms of atomism—namely thermodynamics with notions such as temperature, heat and entropy—was reduced to the fundamental theory based on atomism, namely to classical mechanics via statistical mechanics.

However, there is a bug in this reasoning. The pieces of a broken glass will not come together on their own to form a glass again, because this would require an atypical coordination of the positions and velocities of these pieces. But if this is so, how can there be glasses in the first place that can be broken and thus lead to an increase in entropy? Generally speaking, if there is an increase in entropy, there has to be a low entropy condition in the first place. If we were to apply Boltzmann's statistical reasoning based only on the laws of classical mechanics, which are time reversal invariant, we would get to the conclusion that there is an increase in entropy (uncoordinated positions and velocities of the particles) not only towards the future, but also towards the past. Thus, the present comparatively low entropy, ordered state with glasses on tables, etc., would come out as a spontaneous fluctuation from a past and future disorder. There would be no continuous process that leads towards the present order from a state of even more order (lower entropy). Accordingly, all our records of history that are evidence of such a process would fool us.

To avoid this paradoxical conclusion, it is commonly accepted that one has to trace the rise in entropy back to an initial condition of the universe that is a state of extremely low entropy. This initial condition of the universe is known as the *past hypothesis*. This term was introduced by David Albert (2000, ch. 4). The hypothesis goes back to what Boltzmann (1897) called "assumption A". Thus, if and only if the initial state of the universe is one of highly coordinated initial positions and velocities of the particles, one can then explain the increase in entropy—and thus the *de facto* irreversibility of most of the processes with which we are familiar—in terms of the mentioned statistical reasoning. More precisely, the past hypothesis postulates a boundary condition of the universe in the guise of a low entropy *macro*state, which can still be realized by many different *micro*states of highly coordinated initial particle positions and velocities. Assuming a uniform probability distribution over all these microstates then gets us to the result that the by far vast majority of these microstates leads to an evolution from the low entropy macrostate to higher entropy macrostates (as the by far vast majority of states of broken glasses leads to an evolution in which these glasses remain broken). In these higher entropy macrostates of the universe also appear organisms, which undergo an irreversible process from birth to death. Thus, the past hypothesis is also necessary in order to explain the fact that there is an evolution to living—and finally intelligent—beings in the universe.

The account that includes the past hypothesis as one of its basics is known as the Mentaculus (Loewer 2012), called so after the movie "A serious man", directed by the Coen brothers, in which a character works on a probability map of the universe, which he names "the Mentaculus". This account is a package consisting in (1) the fundamental dynamical laws, (2) the past hypothesis and (3) a uniform probability distribution over all the possible microstates of the universe that realize an initial macrostate that satisfies the past hypothesis. What is missing in this account is spelling out of a primitive ontology. It hence remains an open issue to what the dynamical laws, the past hypothesis and the possible micro-states refer.

One may have the impression that the mentioned bug continues to haunt this account: the past hypothesis amounts to explaining why there are glasses on tables, whose scattered pieces do not come together spontaneously to form a glass again when they are broken, by postulating an initial state of the universe that is as special as were a state in which the scattered pieces of a glass would indeed spontaneously form a glass again. The low entropy initial state of the universe is indeed special in the sense that this explanation postulates one particular type of initial state of the universe instead of applying to whatever state one might take to be conceivable as the initial state of the universe.

However, the low entropy initial state is not atypical. It is dubious to apply a typicality measure to initial conditions of the whole universe, for there is only one universe and only one history of the universe. The physical theory with all its apparatus is there to capture the evolution of that one universe (instead of speculating about other allegedly possible universes). Accordingly, the theory can only be confirmed (or falsified) by observations about the evolution of that one universe. Hence, the reason for conceiving a typicality or probability measure is to make statements about particular systems *within* the universe, namely statements about a large number of repeatable similar situations in which these systems can be, such as a vast number of large series of coin tosses, or a vast number of hot coffees into which cold milk is poured, or a vast number of glasses falling to the floor, etc. Considering all the possible initial conditions in which the particles that form such subsystems of the universe can be, the aim is to obtain statements about what holds for the by far vast majority of initial conditions in such situations, such that probabilistic predictions about their evolution become possible.

Furthermore, Albert (2000) and Loewer (2012) take the past hypothesis also to account for the direction of time. What we take to be the direction of time is reduced to an order in the sequence of the states of motion of the configuration of matter of the universe that is such that this sequence exhibits a low entropy boundary condition. Taking this boundary condition as starting point, there is an order in what occurs in the universe that is directed towards higher entropy states. According to

Albert and Loewer, this order *is* what we call "the direction of time"; it accounts also for the difference in our experience between a past that is fixed (due to lower entropy) and a future that is open (due to higher entropy).

However, an order in the sequence of states of the universe that is imposed by a low entropy boundary condition can also apply to a static universe. In other words, it does not accommodate becoming, that is, the coming into being of these states. One can therefore reject the reduction of the direction of time to the increase in entropy: the past is not fixed because of lower entropy, but because it is past; the future is not open— in the sense that we do not have memory of the future—because of higher entropy, but because it is not there as yet. Hence, one can argue in favour of endorsing the direction of time as a primitive, as done, for instance, by Tim Maudlin (2002).

If one rejects the reduction of the direction of time to the increase in entropy for these reasons, one is, however, not automatically committed to endorsing absolute time. There also is Leibnizean relationalism about time. As mentioned above at the beginning of Sect. 1.3, Leibniz has a strong argument against an ontological commitment to absolute time (i.e. the commitment to surplus structure). But he still defines time as the order of succession, thereby assuming that change *per se* has a direction (that is, independently of whether or not there is a special initial condition such as a low entropy macrostate). The time parameter is introduced in order to be able to measure the change of the relative positions of the objects within the universe. In Leibnizian terms, change goes continuously from one arrangement of distance relations to other such arrangements such that there is a unique order of succession in the configuration of matter of the universe, whatever the initial and the subsequent arrangements may be like—that is, whatever that change is like.

The point of Leibnizean relationalism for present purposes is that change as a primitive that is directed is sufficient to accommodate becoming without having to resort to absolute time and a passage of time. In particular, it is sufficient to accommodate our experience of the future being open in the sense that what there will be in the future is in part up to our free will. We will mention this issue again in Sect. 1.6 and go into it in more detail in Sect. 2.4. To put the argument for the Leibnizean and atomist primitive ontology in a nutshell here, the ontological primitives of permanent matter points individuated by distance relations and perpetual change in these relations strike a perfect balance between Parmenedian being and Heraclitean change.

To sum up, we have identified four ingredients that make up a physical theory in this section and the two preceding ones: (1) a *primitive ontology*, that is, an assumption about what there simply exists in the world; (2) a *dynamical structure*, consisting in a geometry and dynamical parameters that are introduced in terms of their functional role for the evolution of what simply exists in the world, resulting in laws of motion; (3) a procedure how to get *from the laws to probabilities* that allow to make statistical predictions about the outcomes of repeatable processes that are highly sensitive to slight variations in their initial conditions and whose exact initial conditions are not known; (4) an assumption about a *particular initial state of the universe*.

1.5 Beyond Classical Mechanics: Classical Field Theory

There is more to physics than classical mechanics. In particular, there is more to say about interactions than is contained in Newton's force laws. As discussed above in Sect. 1.3 by means of the law of gravitation, this law comes down to conceiving the gravitational acceleration of any particle at a given time as being determined by the positions, velocities and masses of in the last resort all the particles in the universe at that very time modulo the gravitational constant. There hence is correlated particle motion, but no medium propagating in space that establishes this correlation. This is known as "action at a distance". However, Newton himself dismisses action at a distance. In a famous letter to Bentley, he writes:

> It is unconceivable that inanimate brute matter should (without the mediation of something else which is not material) operate upon and affect other matter without mutual contact ... That gravity should be innate, inherent and essential to matter so that one body may act upon another at a distance through a vacuum without the mediation of anything else by and through

which their action or force may be conveyed from one to another is to me so great an absurdity that I believe no man who has in philosophical matters any competent faculty of thinking can ever fall into it. (Letter 406 to Bentley 25 Feb. 1692/3; quoted from Newton 1961, pp. 253–254).

Thus, although Newton is clear about the world being such that there is interaction of objects without there being direct contact between them, he still calls for something in space that mediates the interaction. Nevertheless, his own theory leaves no room for a mediator, since that theory conceives the gravitational interaction as instantaneous—the gravitational acceleration of any object at any given time is fixed by parameters that obtain in the universe at that very time. Consequently, there is no time for something to propagate in space and to mediate the interaction between spatially separated objects.

However, there are more types of interaction than gravitation. There notably is electricity and magnetism, which is also relevant at macroscopic scales. A cogent physical theory of electrodynamics was conceived only in the nineteenth century by James Clerk Maxwell and Hendrik Lorentz, unifying the phenomena of electricity and magnetism. At first glance, this theory looks like a Newtonian force theory, with the force now being the electromagnetic one and the parameter attributed to the particles to which this force can be traced back being their charge. The charge can be positive or negative such that the resulting acceleration is attraction of opposite charges or repulsion of like charges. That notwithstanding, there is a huge difference with Newtonian gravitation: the electromagnetic interaction is retarded instead of instantaneous. That is to say: in order to know the electromagnetic acceleration of a given particle at a given time, one does not have to know the distribution of the charged particles at that very time, but the past distribution of the charged particles. Since the interaction is retarded, there now is the conceptual possibility open to conceive a mediator of that interaction, because there is the time that such a mediator needs to propagate in space. Indeed, such a mediator is introduced in terms of the electromagnetic field. The story is that each charged particle, in virtue of its charge, creates a field around itself that propagates in space with a huge, but finite velocity (i.e. the speed of light). These fields merge into one electromagnetic field. The electromagnetic interaction among the particles takes place through this field as mediator. That is why the interaction is retarded and is considered to be local in the sense of propagating in space with a finite velocity.

This story answers the intuitive demand, voiced by Newton himself, for interaction being such that if two objects interact, something literally propagates from the one object to the other one. But there are strong reasons to doubt that this story is true. In the first place, if each charged particle creates a field around itself or contributes to a common field, then that field acts not only on the other particles, but it also acts back on the very particle that is its source. This self-interaction has the consequence that the field strength becomes infinite at the point where this particle is located. This means that the theory breaks down when one takes the interaction of the particle with the field that it creates or to which it contributes into account. Thus, strictly speaking, the Maxwell-Lorentz theory is mathematically inconsistent: one can use the Maxwell equations to calculate how charged particles influence the electromagnetic field, and one can employ the Lorentz equation to calculate the influence of an incoming, external electromagnetic field on the motion of a given charged particle. But one cannot bring these equations together in a consistent theory.[15]

Furthermore, the electromagnetic field propagates to infinity and thus far beyond where it could influence any particle motion. It therefore looks like a surplus structure if it is taken literally as something that exists in nature. The impression of a surplus structure is reinforced when one asks what physical entity the electromagnetic field could be: Is it a property? Indeed, the most widespread stance to include the electromagnetic field into the ontology is to conceive it as a property, namely as a property of space-time points, given that the field is evaluated at space-time points with field values being attributed to these points. Obviously, this conception implies a commitment to the existence of space-time points over and above the existence of particles and thus a commitment to an absolute space and time. In that vein, what is known as the field argument, formulated by the philosopher Hartry Field (1980, ch. 4, in particular p. 35),

---

[15] See Lazarovici (2018a) for details.

claims that we should admit space-time to the ontology because fields are evaluated at space-time points.

However, field values would be quite odd properties of space-time points. Assuming that space-time points exist, they have metrical properties, making up the geometry of space-time. But one can with good reason maintain that it does not make sense to attribute over and above that electromagnetic field properties to some space-time points, namely to those points where the field value is not zero. It is the ubiquity of gravitation—by contrast to the electromagnetic interaction that concerns only particles with a charge—that makes it possible for Einstein to identify the gravitational field with the metrical field of space-time in the general theory of relativity. But there is no such identification possible as far as the electromagnetic field is concerned.

The other possibility to include the electromagnetic field in the ontology is to conceive it not as a property, but as a substance, namely as some sort of stuff that fills all of space. But what then about the points or regions of space-time where the field value is zero? Is there no field stuff in these regions? Or does the field stuff exist everywhere and merely exerts no force in these regions? More generally speaking, why should there be two sorts of matter, particles and fields, whose interaction is furthermore unclear (as is evident from the mentioned problem of self-interaction)?

However, given the importance of the field concept in physics since the advent of classical electrodynamics, one could also turn the table and argue that matter is fields only so that there are no particles at all, that is, replace a particle monism with a field monism in the ontology of matter.[16] There are two central objections against this move, a metaphysical one and a scientific one, which I already mentioned in Sect. 1.2: the metaphysical objection is that such a view is committed to a—mysterious—bare stuff substratum of matter. On the one hand, the stuff is without structure; one the other hand, it has to admit different degrees of density at different points or regions of space-time as a primitive matter of fact—there is more stuff in some regions of space than in others (whereas in atomism, the distance relations are available to individuate the point particles without the need arising to attribute to them a

---

[16] See Rovelli (1997) for such a view.

primitive stuff substratum or to conceive them as bare particulars). The empirical objection is that, as things stand, there is no precise scientific theory available that gives an account of discrete objects from the atoms in the sense of the chemical elements to molecules and finally to discrete macroscopic objects on the basis of one continuous matter stuff (instead of taking all these things to be composed of elementary particles).

Indeed, all the evidence that we have is one of discrete objects and their motions only. As already noted in Sect. 1.1 when mentioning gravitational waves, all the evidence that we have is change of positions of discrete objects. Fields or waves are employed as means to capture particular forms of motion of discrete objects, such as the patterns in particle motion that are characteristic of electromagnetic attraction and repulsion. Hence, although it may look attractive from the perspective of a first glance at the field concept in physics, there is no serious ontological question of replacing the commitment to particles with a commitment to fields in the ontology of the natural world.[17]

Feynman brings the reservations about the existence of the electromagnetic field to the point in his Nobel Price Lecture:

> You see, if all charges contribute to making a single common field, and if that common field acts back on all the charges, then each charge must act back on itself. Well, that is where the mistake was, there was no field. It was just that when you shook one charge, another would shake later. There was a direct interaction between charges, albeit with a delay. ... Now, this has the attractive feature that it solves both problems at once. First, I can say immediately, I don't let the electron act on itself, I just let this act on that, hence, no self-energy! Secondly, there is not an infinite number of degrees of freedom in the field. There is no field at all. (Feynman 1966, pp. 699–700)

In the terms used in this book, this is to say that the field does not belong to the primitive ontology, but to the dynamical structure. What there exists in nature are the particles. Their motion is correlated in such a way that at least some significant patterns are retarded correlations, some of which can be characterized as electromagnetic attraction and repulsion.

---

[17] See again Lazarovici (2018a) for an elaboration on these arguments.

To represent mathematically this retarded interaction, one introduces the electromagnetic field. But there is no field in nature that literally propagates, mediating the interaction of the particles. So we have retarded interaction, but no mediator of the interaction.

Since, again, fields seem to be concrete physical entities on a par with the particles, it is, like in the case of absolute space and time, helpful to corroborate the stance that refuses to admit them to the ontology if one can vindicate the in principle possibility to formulate the dynamics of classical electromagnetism without employing parameters that attribute field values to space-time points (and, obviously, one thereby avoids also the mathematical problems that come with such fields in the formalism). Indeed, such a mathematical theory was developed by John Wheeler and Richard Feynman (1945). That theory works not only with retarded, but also with advanced direct interaction between charged particles. That is to say, not only the distribution of past charges, but also the distribution of future charges enters into the determination of the electromagnetic acceleration of a given charged particle. Indeed, if the dynamical structure of a physical theory does not conceptualize interaction as instantaneous, it is to be expected that parameters about both the past and the future evolution of the objects figure in the dynamical laws of the theory. Why should one admit in a fundamental physical theory only parameters that refer to the past, thereby breaking the temporal symmetry of the laws? Note that including such parameters into the dynamical structure in order to determine the motion of a given particle does not mean that there is something in nature that literally propagates from the past or from the future.

That notwithstanding, it is evident that there is a problem with the application of any such theory, even if it assumes only retarded interaction: dynamical equations are solved by feeding them with initial conditions. Thus, in Newtonian gravitation, the central initial condition for solving the equation to calculate the acceleration of a particle at a given time is the distribution of the masses at that very time. However, if the interaction is retarded (and possibly also advanced) instead of instantaneous, it is not clear what exactly are the initial conditions for solving the dynamical equations, that is, exactly what past (and possibly also future) trajectories of charged particles have to be taken into account as initial condition. Here shows the mathematical elegance of working with fields in the formalism up: they allow for the precise formulation of initial value problems, since all that is relevant for the calculation of the electromagnetic acceleration of a given charged particle are the field values in its immediate spatio-temporal vicinity. However, it is also the case that in order to know these field values, one has to know the trajectories of the charged particles in the past that determine these field values. Consequently, in the end, these field values are not independent degrees of freedom; the problem how to fix initial values arises also in a dynamical structure with fields as recently shown by Dirk-André Deckert and Vera Hartenstein (2016).[18] In short, in any theory of retarded interaction, knowledge of the past is required to calculate the present particle motion.

In sum, one has to distinguish between the following two issues: is physical interaction instantaneous, or is it retarded (and possibly also advanced)? Is physical interaction direct, or is there a mediator that literally propagates in space (so that it belongs to the ontology instead of merely being a mathematical means for calculations)? Newton's call for such a mediator in the quotation at the beginning of this section can with good reason be dismissed as being based on an anthropocentric intuition of interaction having to proceed by direct contact. This is the basic reason for scepticism about an ontological commitment to fields. This reason then is corroborated by the mathematical as well as philosophical problems to which the ontological commitment to fields as mediators of interactions leads. Consider how Bertrand Russell describes Newtonian gravitation:

> In the motions of mutually gravitating bodies, there is nothing that can be called a cause and nothing that can be called an effect; there is merely a formula. Certain differential equations can be found, which hold at every instant for every particle of the system, and which, given the configuration and velocities at one instant, or the configurations at two instants, render the configuration at any other earlier or later instant theoretically calculable. That is to say, the configuration at any instant is a function of that instant and the configurations at two given instants. This statement holds throughout physics, and not only in the special case of gravitation. (Russell 1912, p. 14)

---

[18] For an assessment of the philosophical consequences, see, over and above Lazarovici (2018a), also Hartenstein and Hubert (2019).

The notion of causation that Russell has in mind here is the one of something literally propagating in space. As he points out, such an intuition is misplaced throughout the whole of physics.

1.6   From Field Theory to Relativity Physics

If the particle interaction is retarded (and possibly also advanced) as in electrodynamics, the dynamics does not rely on information about what happens elsewhere in space at the same time. There then is a maximal velocity for the propagation of effects. This is the speed of light. By contrast, in Newtonian mechanics, arbitrarily high velocities are allowed. Furthermore, since the interaction is instantaneous, there has to be a well-defined simultaneity of events, which is absolute in the sense that it is independent of the choice of a reference frame. The reason is that what happens at a point in space is determined by what there is elsewhere in space at the same time.

Absolute simultaneity in this sense does not require the commitment to an absolute scale: also in Leibnizian relationalism, there are absolute instantaneous configurations that are defined by ratios of distances (and angles, as in shape dynamics), whereby all magnitudes are scale-invariant. Consequently, absolute simultaneity of events does not commit us to absolute time in the sense of an external time parameter. More precisely, in virtue of the well-defined, objective simultaneity of events, there is a unique temporal order of all the events in the universe: any two events are either simultaneous, or one of the two events occurs earlier than the other one. But in the absence of an external time parameter, it makes no sense to compare different possible histories of the universe and to ask what happens according to these histories at a certain time, that is, to ask, given a certain arrangement of matter in the actual history of the universe, what would be the corresponding arrangement of matter in a possible alternative history of the universe. In relational mechanics, there is no determinate time that would be the same for different possible evolutions of the universe.

In Newtonian mechanics, there is absolute space and time, and any interaction between particles is instantaneous. Consequently, Newtonian mechanics needs the objective simultaneity of events. However, any measurable velocity can be relative to a particular reference frame, and arbitrarily high velocities can be allowed. By contrast, if the interaction is retarded and hence there is a maximal velocity for the propagation of effects, this velocity is absolute in the sense that it is the same in all reference frames (i.e. the speed of light). There then is no need for objective simultaneity any more. Any simultaneity of two or more events can then be taken to be relative to a particular reference frame. In this case, there is no unique (that is, frame-independent) temporal order of the events in the universe. Hence, in this case, absolute space and absolute time go. However, an absolute space-time remains.

More precisely, construed in this vein, the swift from a dynamical structure that conceives interaction as instantaneous (Newtonian gravitation) to a dynamical structure that conceives interaction as retarded (electrodynamics) implies that one also has to change the geometry of space and time. That change is carried out by Albert Einstein in the *special theory of relativity* in 1905. The term "relativity" refers to the fact that in this theory, simultaneity—and thereby the temporal order of events—is relative to the choice of a particular reference frame. In fact, however, there is not more relative in relativity physics than in Newtonian physics; it is just that other things are relative and absolute. In Newtonian physics, any measurable velocity is relative to a reference frame, and simultaneity is absolute. In relativity physics, there is one absolute velocity (i.e. the one of light), and simultaneity is relative to a reference frame.

The light cone structure is central for the geometry of relativistic space-time.[19] For any event e occurring at a space-time point, there is a forward or future light cone consisting in all and only those events that can be reached from e by a signal that propagates maximally with the speed of light; and there is a backward or past light cone consisting in all and only those events from which e can be reached by a signal that propagates maximally with the speed of light. This implies that there are events outside the past and future light cones of e. No temporal order is defined for these events, unless one chooses a particular reference frame. That is why simultaneity is relative to a reference frame.

---

[19] For an excellent exposition, see Maudlin (2012, chs. 4–5).

When passing from one reference frame to another one, it is necessary to pay heed to the speed of light being the maximal velocity for the propagation of effects. That is why one can no longer use the Galilei transformations that are valid in Newtonian mechanics. One has instead to use the Lorentz transformations when switching from one reference frame to another one. The latter transform both the three-dimensional spatial distances and the one-dimensional temporal intervals between events when passing from one reference frame to another one. But they leave the four-dimensional space-temporal interval between any two events unchanged. That is why space and time are taken to be united in a four-dimensional space-time in special relativity theory. That space-time is as absolute as are Newtonian space and time: it is taken to simply exist, with the configuration of matter being embedded in it.

The *general theory of relativity*, achieved by Einstein in 1915, adds to the special theory a field theory of gravitation based on four-dimensional space-time. Consequently, gravitation is no longer conceived as action at a distance, as in Newtonian mechanics. However, there is no specific field on space-time for the gravitational interaction, as there is a specific field for the electromagnetic interaction. The gravitational field is identical with the metrical field of space-time, that is, the field that defines the geometry of four-dimensional space-time. Consequent upon this identification, the geometry of space-time is no longer Euclidean, but Riemannian. Space-time is represented as being curved, with its curvature being influenced by the distribution of masses; the curvature of space-time, in turn, influences the motion of the masses. That is the essence of Einstein's conception of gravitation as local interaction by contrast to action at a distance.

However, the gravitational field is not determined by the masses as its source, as the electromagnetic field is determined by the charges as its source.[20] This is so because the gravitational field is identical with the metrical field. The distance between the physical objects cannot be taken

---

[20] There is no reason to admit fundamentally source-free fields in electrodynamics, although for practical purposes, one calculates with external fields without considering their sources. As mentioned in the previous section, fields in classical electrodynamics are in the end not independent degrees of freedom. See again Deckert and Hartenstein (2016) as well as Hartenstein and Hubert (2019).

to be determined by parameters that are attributed to these objects taken individually (such as their mass). It is only when one fixes a metric of space-time (curvature) as initial condition and an initial distribution of the masses that the further evolution of both the metrical field of space-time and the motion of the masses is determined by Einstein's field equations. Furthermore, the identification of the gravitational field with the metrical field confirms the ubiquity of gravitation as an interaction that concerns all the material objects, whereas the electromagnetic interaction concerns only the charged particles.

As far as we know, any model or solution of the Einstein field equations that is in the position to describe the actual universe allows for a foliation of four-dimensional space-time into three-dimensional spatial hypersurfaces that are ordered in one-dimensional time. This means that the trajectory of any material object crosses any such hypersurface only once; there are no closed timelike curves. But there is no evidence that would allow us to detect in the universe one privileged foliation of four-dimensional space-time into three-dimensional hypersurfaces that are ordered in one-dimensional time. Even if one talks in terms of a cosmic time and indicates an age of the universe from the big bang on, this does not amount to detecting a universally privileged reference frame or foliation of space-time. In principle, many different reference frames or foliations of space-time can be employed in the sense that they all yield an empirically correct description of the evolution of the universe as far as we know it. In other words, the space-time geometry of the general theory of relativity is refoliation invariant.

For this reason, the primitive ontology of Leibnizian relationalism seems to be tied to pre-Einsteinian physics in that it conceives distance relations—and thus spatial relations—as individuating the particles and unifying the world and sets out a dynamics for the change of these relations. Since these relations are not invariant in relativity physics, it may therefore seem that consequent upon the advent of the special and the general theory of relativity, one has to switch in the primitive ontology from primitive spatial distance relations and their change to primitive spatio-temporal relations. However, at least three reservations speak against such a switch.

(1) In the first place, interaction theories in terms of fields have internal consistency problems, as is evident from the electron self-interaction problem in classical electrodynamics mentioned in the previous section. In general relativity theory, one represents matter as a continuous fluid, since the theory is usually applied at large scales. Nevertheless, this is only a coarse-grained description. Matter is composed of point particles. If one considers point particles, the problem of a point particle influencing a field and that field, as influenced by the particle, reacting back on that very particle, is not resolved in general relativity theory. This problem, together with the other ones discussed in the preceding section, is a warning post against granting fields the status of belonging to the primitive ontology, including the metrical field of general relativity theory.

(2) Furthermore, the geometry of special and general relativity theory does not turn time into the fourth dimension of space. If time were just a further dimension of space, the theory could not accommodate the local temporal order of events that we observe. There is an objective difference between the events that take place within the light cones of a given event and the events that occur outside the light cones. The geometrical structure of light cones enables us to define a local temporal order for each event in the sense of a set of events that are in its past, whatever reference frame one chooses, and a set of events that are in its future, whatever reference frame one chooses. In other words, the spatio-temporal relations allow to individuate the events in such a way that there is an absolute (i.e. not frame-relative) difference between events that are separated by a timelike spatio-temporal interval from each other (i.e. that are within each other's light cones, being either past or present) and events that are separated by a spacelike interval from each other (i.e. that are outside each other's light cones with no temporal order defined).

However, the concern is that an only local temporal order of events is not sufficient to accommodate change. When one switches from primitive spatial to primitive spatio-temporal relations in the ontology, particles and their trajectories can be construed as sequences of point particle events that form continuous lines (worldlines), which are individuated by the spatio-temporal relations between them. But these relations do not change: they are spatio-temporal. That is why the ontology based on primitive spatio-temporal relations is known as *block universe*: all the events throughout the history of the universe exist at once, being separated by spatio-temporal intervals.

Of course, when we cut three-dimensional slices through the four-dimensional space-time geometry and compare them, we can define change in terms of the differences between such slices. But this change concerns only an arbitrary description and not what exists in nature, namely the four-dimensional block. The objection therefore is that the block universe ontology does not have the means to distinguish between *variation* and *change*. Variation within a configuration consists in the different relative distances in which the objects that build up the configuration stand to each other (whatever may be the dimensionality of these distances as represented in a certain geometry). Change consists in the fact that these very distances become different, from which a temporal development of the configuration results. If there is no objective temporal development of the configuration, then there is also no change of the distances in contrast to their variation within a stationary configuration. Peter Geach (1965, p. 323) highlights this objection in terms of a poker *being* hot at one end and cold at the other one (variation within a given configuration) by contrast to the poker *changing* from cold to hot (change for which there is no place in the four-dimensional block).

There is variation in the block universe as given by the different spatio-temporal relations that individuate the point events and point particle trajectories (worldlines). Furthermore, the block universe can be structured in such a way that there is a low entropy state at one end and a state of thermal equilibrium at the other end. But, still, this is on a par with variation in a given configuration of point particles that are individuated by the distances between them, only that the distances now are four-dimensional with a fully-fledged metric. For instance, a given configuration of point particles can be such that on the one end, particles are concentrated in a dense manner, whereas at the other end, particles are far apart from one another. There is no change here from a low entropy state to a state of thermal equilibrium in exactly the same way as there is no change in the mentioned configuration from densely concentrated particles to particles with large distances between them. This is the—correct—root of the objection that is crudely put in terms of time being treated as a further dimension of space in special and general relativity (if the conclusion of the block universe ontology is drawn from these theories).

One can make use of the structure in the block universe of a low entropy state at one end and a state of thermal equilibrium at the other end to simply stipulate that this structure defines a time axis and thereby change from the one to the other state. However, such a stipulation does not remove the mentioned objection. The issue is whether what is admitted in the primitive ontology is such that it entitles one to draw a distinction between variation and change. The objection is that there is nothing in the ontology that licenses drawing this distinction, since the ontology contains only distances between point-events on continuous worldlines, with these distances being spelled out as four-dimensional intervals that all exist at once. Hence, there is variation, but no change of anything.

To stress again, the point at issue here is not the controversy in metaphysics between eternalism—according to which everything that occurs in time simply exists—and presentism, according to which only the present exists. On Leibnizean relationalism, change is eternal. What exists is both the configuration of point particles as individuated by the distances among them and the change in this configuration, that is the change of the distance relations. One can therefore take the whole change to simply exist, or one can maintain that only the present arrangement of the point particles exists, which is in continuous change. That is to say, Leibnizean relationalism can go both with eternalism and presentism, properly formulated to fit this primitive ontology. The point at issue is whether the primitive ontology entitles one to draw a distinction between variation and change. This is so in Leibnizean relationalism by definition, whereas the entitlement to drawing this distinction fades away if one switches in the primitive ontology from primitive spatial to primitive spatio-temporal relations: these latter can vary, but not change.

By way of consequence, the block universe ontology cannot include becoming. Hermann Weyl is famous for saying:

> The objective world simply *is*, it does not *happen*. Only to the gaze of my consciousness, crawling upward along the life line of my body, does a section of this world come to life as a fleeting image in space which continuously changes in time. (Weyl 1949, p. 116)

Change and becoming hence are relegated to the consciousness of persons. They are not part of the physical world. One thereby provokes a conflict on the issue of time and change between the scientific image of the world, conceived in terms of a block universe, and the manifest image of the world.

(3) Even leaving these objections aside, the advent of relativity physics does not oblige us to switch in the ontology from primitive spatial to primitive spatio-temporal relations. The situation here is the same as in the case of absolute space and time in Newtonian mechanics and the case of fields in addition to particles in electrodynamics. Scientific realism as regards the physics does not force us to subscribe to an ontological commitment to these things, although they look like concrete physical things on a par with the particles. To substantiate this claim, again, it is helpful to be in the position to point out how the physics can be formulated without using these things in its dynamical structure.

The relevant point in that respect is to realize that the general theory of relativity is refoliation invariant: the theory does not pick out an instantaneous configuration of matter in terms of a three-dimensional geometry through a privileged foliation of space-time into three-dimensional spatial hypersurfaces that are ordered in one-dimensional time. However, it is not scale invariant. The spatio-temporal intervals between point-events are absolute magnitudes in distinction to magnitudes for which only their ratios are relevant (as in the case of the distance relations that define an instantaneous configuration). Hence, if one switches in the ontology from distance relations as the world-making relations to spatio-temporal relations, one has to presuppose that there are absolute spatio-temporal intervals between non-simultaneous point events as a primitive fact. Consequently, there is an absolute scale that defines the spatio-temporal intervals between events. Otherwise, one would not have the light cone structure of space-time at one's disposal and would thereby lose the above mentioned local temporal order of events.

This fact suggests that one can either have a theory of gravitation that is refoliation invariant, but that has to buy into absolute scale in the guise of absolute space-time intervals; or one can have such a theory that is scale-invariant, but then relies on successions of instantaneous configurations and thereby on absolute simultaneity. In other words, one can have either refoliation invariance or scale invariance, but not both. Indeed, it is possible to develop Barbour's relational mechanics or shape dynamics as the geometry of a relational theory of gravitation that is an alternative to the geometry of an absolute space-time in general relativity.[21] In this case, one requires successive well-defined three-dimensional geometrical configurations and thus absolute simultaneity—in other words, an absolute distinction between space as the order of what coexists and time as the order of succession. But one employs only scale-invariant quantities. Again, the latter is an in principle possibility, not a practical advice for carrying out calculations. Nonetheless, the issue of two different dynamical structures for gravitation cannot be settled by observation: both yield the same particle trajectories given appropriate restrictions (such as the restriction in general relativity theory to four-dimensional geometries that admit a foliation into three-dimensional spatial hypersurfaces that are ordered in one-dimensional time). One can neither observe absolute scales nor absolute simultaneity. The one framework works with the theoretical postulate of scale-dependent, spatio-temporal intervals, the other one with the theoretical postulate of successive global, three-dimensional geometrical configurations to achieve a dynamics of gravitation.

This fact confirms again that the primitive ontology is one thing, and the dynamical structure another thing. There is no need to abandon the distinction between variation within a configuration and change of the configuration and to provoke a conflict between the scientific and the manifest image of the world on the subject of temporal becoming consequent upon the advent of the dynamical structure encoded in the general theory of relativity.

1.7   From Statistical Mechanics to Quantum Mechanics

For Einstein, the story that physics tells about the world could have been complete with the treatment of gravitation as local interaction in the general theory of relativity. However, there then came the transition from

---

[21] See Gomes et al. (2011), Gomes and Koslowski (2013), Mercati (2018, ch. 7). See furthermore Gryb and Thébault (2016, in particular pp. 692–697) for a philosophical discussion.

classical to quantum mechanics in the 1920s. One can approach this transition from statistical mechanics, since quantum mechanics allows in general only statistical predictions of measurement outcomes. Suppose that we live in a universe of classical mechanics in which planetary motion or the throw of a stone on Earth are not available as paradigmatic examples of applying the deterministic laws to obtain deterministic predictions. Instead, all the processes are like the coin toss: only statistical predictions are available. A world of classical mechanics can be like this. The deterministic laws of classical mechanics are of use also in such a world. First of all, although the laws cannot be employed to generate deterministic predictions, they still are the basis from which the statistical predictions are derived. Moreover, they answer the question of what happens with an individual coin between its release and its arrival on the soil. It would be absurd to maintain that the coin has no trajectory, that it is somehow smeared out over all its possible trajectories, or even that it vanishes after its release and somehow comes back into existence upon its detection on the soil, only because it is not possible to calculate and predict its trajectory.

As far as the status of probabilities is concerned, in contrast to a widespread opinion, the world that quantum physics describes can indeed be construed as being like a classical world of coin tosses. Moreover, it cannot only be thus construed, but in doing so one also resolves the conceptual puzzles of quantum physics, as will become evident in this section. That notwithstanding, there are a number of important differences between classical and quantum mechanics. In the first place, in classical statistical mechanics, the limitation on the knowledge of the exact initial conditions is only a practical one: it is practically impossible to measure the initial positions and velocities of the particles with such a precision as to be able to calculate the trajectory of a particular tossed coin.

In quantum mechanics, this limitation is a principled one, as expressed by the Heisenberg uncertainty relations: it is in principle not possible to measure both the position and the velocity (momentum) of a particle with arbitrary precision. Given that the evolution of the particles is in general highly sensitive to slight variations in the initial conditions in quantum physics, it then follows that there are in general only statistical predictions of measurement outcome distributions possible. Nonetheless, this still is a limitation only on the accessibility of quantum objects. There is nothing in the Heisenberg uncertainty relations *per se* that could warrant the conclusion that quantum objects do not have a precise position and velocity. This would be an ontological conclusion that cannot follow from an epistemic premise about limits on the accessibility of quantum objects alone. In other words, everybody agrees that the Heisenberg uncertainty relations put a principled limit on our ability to measure initial conditions. By contrast, attributing an ontological significance to them is a matter of controversy and requires in any case further premises.

Generally speaking, it is not astonishing that there is a limitation on our knowledge of particular matters of fact. For one particle configuration, such as a macroscopic object, to contain information about the positions of the particles in other configurations, there must be a correlation between the configurations in question, which has to be reliable in the sense of being stable and reproducible. This applies in particular to correlations between particle configurations in human brains and particles outside the brains, assuming that all the perceptual knowledge that persons acquire passes through their brains (in the sense of a necessary, not automatically a sufficient condition). Hence, for reasons stemming from the way in which we acquire knowledge about the natural world, the possibility of deterministic predictions turns out to be rather a lucky case and the possibility of statistical predictions the expected case, although the laws can be fully deterministic. In that sense, classical mechanics is an idealization, and quantum mechanics brings out that limitation on our knowledge. That notwithstanding, there is no question of an *a priori* deduction of the Heisenberg uncertainty relations from general conditions of our knowledge. It is just that some principled limit on our knowledge of particular matters of fact—such as initial conditions of physical objects—is to be expected.

Consider the famous double slits experiment. This is an experimental arrangement in which particles are emitted from a source, then pass through a barrier with two slits and are finally registered on a screen. The particles can be sent one by one through this device. When both slits are open and the experiment is run with many particles, by contrast to what would be expected from classical mechanics, the typical distribution of the particles at the end of the experiment manifests an interference pattern. Such a pattern, however, does not show up if only one slit is open. As in the classical coin toss case, we cannot predict where an individual particle will hit the screen. Hence, if both slits are open, one cannot predict through which slit an individual particle runs. The only prediction that we can make is that if the experiment is run with many particles, the particle distribution on the screen at the end of the experiment will typically manifest a particular interference pattern. This statistical prediction is achieved by means of representing the evolution of the particles in terms of a wave-function that behaves like a wave in that it undergoes superpositions and interference in the mathematical space in which it is defined.

However, the question is what happens with each individual particle. How does it get from the source of the experiment to the screen? The textbook presentations of quantum mechanics do not answer this question. The formalism of quantum mechanics given in the textbooks is an algorithm to calculate measurement outcome statistics. It is not physics in the sense of a theory that tells us what there is in the physical world and how what there is behaves (or that enables inferences that answer these questions), as pointed out most recently by Maudlin (2019, introduction). Taking it to be physics in this sense runs into the well-known paradoxes such as the measurement problem, which is nicely illustrated by Schrödinger's cat.[22] The cat is represented as being in a superposition of being *both* alive *and* dead given the evolution of the wave-function according to the Schrödinger equation. However, when measured, it is found to be *either* alive *or* dead. The issue hence is how the theory accounts for determinate measurement outcomes and how it conceives the evolution of the individual physical objects, such as the individual particles sent through the double slits experiment, or the particles in the Schrödinger cat experiment.

Given the state of the art in foundational research in quantum physics, there are two types of options that one can pursue in order to achieve a quantum theory that answers these questions.[23] The one type of option is to take the quantum mechanical wave-function and its evolution as the

---

[22] See Maudlin (1995) for a precise formulation of the measurement problem.
[23] See Wallace (2008) for a general overview.

guide to physical reality and to make intelligible how one can live with the ensuing consequences. The most notable consequence is that everything that is possible in the dynamical evolution of the wave-function according to the—linear and deterministic—Schrödinger equation then is taken to become in fact real. Thus, the very same cat in the Schrödinger cat thought experiment is both alive and dead, albeit in different branches of the universe that do not interfere with one another. Hence, according to this option, many branches of the universe—known as "many worlds"—exist in parallel. This option goes back to Hugh Everett (1957). Prominent contemporary elaborations of taking the wave-function as the guide to physical reality are the books by David Wallace (2012) and, in a different manner, David Albert (2015, chs. 6–7). The challenge then is to explain measurement outcomes and records on this basis. Furthermore, it is an open issue what the status of probabilities could be if everything that is possible according to the theory comes in fact into existence in a branch of the universe with all these branches being real. Consequently, there is no fact of the matter in which branch a person and her experience is situated: each person exists in many branches and lives all her possible futures in different branches, although these futures do not interfere with one another.

The other type of option is to grant the wave-function only a dynamical status in representing the evolution of a primitive ontology of objects in ordinary physical space, which, consequently, do not enter into superpositions. This primitive ontology then has to be represented in the formalism by means of a parameter in addition to the wave-function. In other words, the quantum state or wave-function then is not the primitive ontology in the wide sense in which this term is used in this book as standing for whatever is the basic ontology according to a theory. It is only a dynamical parameter. There are different possibilities for setting out a primitive ontology of quantum mechanics in physical space as well as for the dynamics of the wave-function. Charles Wesley Cowan and Roderich Tumulka (2016) have shown that in any such theory, the primitive ontology—that is, the distribution of matter in physical space—is not fully accessible to an observer in the universe; otherwise, paradoxes would ensue such as the ones resulting from the possibility to send signals with superluminal velocity.

Hence, if one admits a configuration of matter in physical space—accounting for, among other things, determinate measurement outcomes—, it follows in any case that there is a limit to the epistemic accessibility of that configuration. To employ a term that used to be widespread, such a configuration is in any case a "hidden variable"—independently of whether or not in the dynamics, one endorses jumps of the wave-function in measurement situations that eliminate the superpositions in the wave-function and are therefore known as collapse of the wave-function, and thus independently of whether or not the dynamics is indeterministic (collapse) or deterministic. In other words, going for a dynamics of wave-function collapse cannot avoid the commitment to what used to be called "hidden variables".

Contrary to what many discussions of quantum physics suggest, the situation therefore is this one: if one wants to have a theory that answers the above mentioned questions, one has to go either for "many worlds" or for "hidden variables". There is no other option. That notwithstanding, both the terms "many worlds" and "hidden variables" are misleading: "many worlds" are many branches of the one universe, and "hidden variables" are what becomes evident in the measurement outcomes, namely the configuration of matter in space (whereas the wave-function, if it exists at all, never becomes evident). Thus, the motivation for admitting a primitive ontology of a configuration of matter in space as variable besides the wave-function is to account for measurement outcomes and our observations in general, thereby avoiding any sort of a measurement problem.

Any formalism of non-relativistic quantum mechanics works with the assumption of a determinate number of permanent point particles. Accordingly, the first and to this day most widespread quantum theory with a primitive ontology of matter in space is the one that bases itself on an ontology of point particles that are characterized only by their relative positions. Such a theory was originally proposed by Louis de Broglie (1928), then developed by David Bohm (1952) and promoted by John Bell (2004, in particular chs. 4, 7 and 17); it is today known as Bohmian mechanics.[24] This theory describes the evolution of the particle positions

---

[24] See Dürr et al. (2013). See Goldstein (2017) for an overview, Bricmont (2016) for a recent elaborate defense and Dürr and Teufel (2009) for a textbook presentation.

by a law that is known as the guiding equation and that is such that, given the particle positions at any time t as input, the law fixes the velocities of the particles at t by means of the wave-function as output. That law is deterministic.

This fact makes this theory particularly relevant for the purpose of this book: it shows that there is no quick way of bringing human freedom together with physics by referring to quantum mechanics being allegedly an indeterministic, fundamental physical theory. Quantum mechanics can be cast as a fully deterministic theory while still being committed only to one, always determinate configuration of matter in physical space (that is, in order to retain determinism, one does not have to resort to "many worlds" in which each possible configuration is realized). Any attempt to bring freedom together with physics has to stand up to scrutiny also in the light of such a theory.[25]

Over and above the guiding equation that fixes the particle trajectories by means of the wave-function, Bohmian mechanics uses the Schrödinger equation to describe the evolution of the wave-function itself. This equation can also be employed to calculate probabilities for measurement outcomes. To do so, one needs a probability measure. In Bohmian mechanics, this measure is known as quantum equilibrium, from which then follows the rule of the textbooks (i.e. Born's rule) to calculate measurement outcome statistics for ensembles of quantum objects that are prepared under the same conditions. Hence, on Bohmian mechanics, the probabilities in quantum mechanics have exactly the same status as in classical, statistical mechanics: they are derived from deterministic laws via a probability measure.[26]

This is different in a quantum theory that includes the collapse of the wave-function in its dynamical laws, when the collapse of the wave-function is conceived as an irreducibly stochastic process. The most elaborate such dynamics is the one proposed by Ghirardi et al. (1986), which is known as GRW theory.[27] In this theory, probabilities are fundamental. Nevertheless, this theory also needs a law that links the evolution of the

---

[25] See also Loewer (1996).
[26] See Dürr et al. (2013, ch. 2) for details.
[27] See already Gisin (1984) for a forerunner.

wave-function with the primitive ontology of a configuration of matter in physical space and its evolution.[28]

There are two different proposals for a primitive ontology of matter in space linked with the GRW formalism of wave-function collapse. The one proposal conceives single events in space-time, known as flashes, and goes back to Bell (2004, ch. 22) (the term "flash" was introduced by Tumulka 2006, p. 826). Whenever—and only when—the wave-function collapses in the mathematical space on which it is defined, a flash-event occurs at a point in physical space. These single events are all there is in space-time. They can be considered as Bohmian particles deprived of their trajectories when they are not observable for us due to our ignorance of the exact initial conditions. Thus, in the double slits experiment, there is one flash at the source and one flash on the screen at the end of the experiment, but nothing in between, in particular no trajectory that links these flashes (so that the question through which slit the particle passes does not even arise). The flash ontology is therefore committed to an absolute background space-time in which the flashes occur and that is what unites them so that they are in one world. Furthermore, the flashes are bare particulars: the difference between a space-time point being empty and a flash occurring at it (or its being flashy so to speak) is primitive. This makes the flash-events ontology unattractive from a metaphysical point of view. As regards the physics, apart from reservations that one may have about the GRW law, one may wonder whether macroscopic objects and what appears to be their continuous existence can convincingly be accounted for in terms of "galaxies of flashes" as Bell (2004, p. 205) put it.[29]

The other proposal is the one of a continuous matter density field in space and was formulated by Ghirardi himself (Ghirardi et al. 1995). As Valia Allori et al. (2014, pp. 331–332) put it, the matter density field "does not *ipso facto* have any such properties as mass or charge; it can only assume various levels of density". This proposal is therefore hit by the mentioned metaphysical reservations against an ontology of matter as one continuous stuff that has different degrees of density in different

---

[28] See e.g. Cowan and Tumulka (2016).
[29] See Maudlin (2011, p. 258 and 2019, pp. 113–115) for this objection.

regions of space as a primitive matter of fact (see Sects. 1.2 and 1.5). That notwithstanding, the GRW matter density field ontology is the only concrete and physically precise proposal at hand of such a view of matter and a dynamics to account for the observable phenomena.

However, if compared with the flash ontology, the matter density field ontology turns out to be a surplus structure, since on the GRW dynamics, all the observable phenomena correspond to wave-function collapse events. There hence is no need to admit a continuous matter field over and above the point events—the flashes—in physical space. This surplus structure comes again with a drawback, namely in this case the doubt whether the GRW matter density field ontology provides at the end of the day a solution to the measurement problem: on this field or wave ontology of matter, there are superpositions occurring in fact also in physical space, albeit only in a negligible manner. For instance, if in the Schrödinger cat experiment the outcome is that the cat is alive, there nonetheless is a tiny dead cat superposed with the live cat in physical space on this ontology. The problem is that the low-density cat seems to be just as cat-like (in terms of shape, behaviour, etc.) as the high-density cat, so that there are in fact two cat-shapes in the matter density field, one with a high and another one with a low density.[30]

In comparison to the proposals of a matter density field or flash-events and the corresponding dynamics of wave-function collapse, the Bohmian particle ontology and dynamics arguably offers the better solution to the measurement problem.[31] In brief, if quantum particles also move on trajectories, it is no puzzle why they are always at precise locations when measured. There is no mystery of how they get there—that is, how they come to be concentrated at determinate positions. This is so independently of the fact that we cannot predict these trajectories in the individual case because of a principled ignorance of their exact initial conditions. By the same token, Schrödinger's cat always is either in a particle configuration of a live cat or in a particle configuration of a dead.

---

[30] See e.g. Maudlin (2010, pp. 135–138 and 2019, pp. 117–121) for this objection. See Esfeld (2014a) for a detailed assessment of these proposals.

[31] See again Esfeld (2014a) for a detailed argument.

It is only that we may not know the real particle configuration without observing it.

In general, when we privilege the Bohmian formulation of quantum mechanics known as Bohmian mechanics in this book, this is not so for the a priori reason that this formulation fits best with the primitive ontology of atomism as spelled out in Sect. 1.2, but for the empirical reason that it provides the best solution to the measurement problem and the methodological reason that any discussion of the relationship between free will and physics has to be probed at the face of the availability of a deterministic formulation of quantum mechanics that is committed only to one, always determinate configuration of matter in physical space.

All measurement outcomes come down to observations and records of the relative positions of discrete objects. Notably John Bell (2004, p. 166) stresses this point in saying, as already quoted in part in Sect. 1.1

> … in physics the only observations we must consider are position observations, if only the positions of instrument pointers. It is a great merit of the de Broglie–Bohm picture to force us to consider this fact.

Indeed, in Bohmian mechanics, the operators or observables of textbook quantum mechanics are construed as mathematical means to make statistical predictions about how the particles move in certain experimental contexts. This applies also to spin. Hence, they are derived from position being the only parameter that defines the particles and the law for the change of the particle positions. By way of consequence, operators or observables are not properties of anything, not even contextual properties of measurement situations.[32] Again, Bell (2004, p. 35) brings this point out in saying, "The electron need not turn out to be a small spinning yellow sphere". Spin is not an intrinsic property of physical objects, but a certain pattern of motion that shows up in certain experimental setups. A similar remark applies to the GRW matter density field and flash-events ontologies: they also privilege position as the only physical property and construe the operators as means to calculate the evolution of the field density or the flash positions.

---

[32] See Lazarovici et al. (2018).

Consequently, all the parameters apart from position are situated on the level of the wave-function and are employed to determine the evolution of the configuration of matter in physical space. The wave-function is the central dynamical parameter in fixing that evolution. It can be conceived as being part of a nomological machinery that, given the configuration of matter as input, yields its evolution as output. Thus, in Bohmian mechanics, given the particle positions as input, one gets their velocities and thus their trajectories as output. Hence, in contrast to classical mechanics, the initial condition is not the positions and the velocities of the particles, but their positions only, plus the initial wave-function that is attributed to the particles. The wave-function is not a classical force or field. It is defined on the mathematical space known as configuration space. For N particles, configuration space has $3N$ dimensions, namely one dimension for the position coordinate of each particle in the three directions of physical space. Thus, for a configuration of 6 particles such as the one depicted in Fig. 1.1 in Sect. 1.3, the configuration space has already 18 dimensions. Each point of configuration space thereby represents a possible configuration of the particles in three-dimensional, physical space.

The wave-function hence is a field on configuration space. It is like a wave in that its evolution can undergo superpositions. However, the question is what the wave-function and its evolution on configuration space represent. If the wave-function is a dynamical parameter that is there to fix the evolution of the particle positions in physical space, there is no question of there being superpositions of anything in physical space. The particles always move on determinate trajectories. But these are often not classical trajectories. If we consider an ensemble of particles as in the double slits experiment and ignore the individual particle trajectories, the final particle position distribution on the screen manifests an interference pattern that looks as if a wave passed through the two slits, went into superposition and underwent interference. However, there was no wave passing through the slits, but individual particles moving on non-classical trajectories such that an ensemble of them gives rise to the interference pattern—otherwise, the measurement outcome of a particle distribution with each particle having a precise position (i.e. becoming manifest as a precise dot on the screen) could not be explained.

Moreover, if we consider a system that consists of several particles, its wave-function not only undergoes superpositions, but can be—and generally is—entangled. This means that the wave-function represents the evolution of the particle positions as being bound together, given that it is situated in configuration space with each point of that space representing a possible particle configuration in physical space. More precisely, given the positions of the particles in a configuration at a time t, the velocity that any particle in that configuration acquires at t depends on the positions of all the other particles in the configuration. That dependency is mediated by the wave-function. It is known as quantum non-locality.

This non-locality is already evident in the double slits experiment. The trajectory of any particle after having passed one slit depends on whether or not the other slit is open. The particle configuration that has to be taken into account here is made up not only of the particles that go through the slits but includes also the particles that compose the whole experimental device. This is the explanation why, when both slits are open, the ensemble of the particle trajectories typically manifests an interference pattern; but there is no interference pattern when only one slit is open when the particles pass.

Quantum non-locality has been established by Bell's theorem.[33] Consider again a classical field theory like electrodynamics: all that is relevant for the evolution of a particle that is situated at a point in space-time is contained in the past light cone of that point. What happens outside the past light cone is irrelevant for its evolution. Hence, when calculating the probabilities for what happens at any given space-time point, parameters that are situated outside the past light cone can be neglected. Such parameters cannot alter these probabilities.

However, Bell's theorem proves that, for certain observables, one cannot obtain the probabilities of quantum mechanics for measurement outcomes performed at a certain space-time region if one does not take into account parameters that refer to what happens outside the past light cone of that region, such as fixing parameters to be measured and measure-

---

[33] See Bell (2004, notably chs. 2, 7, and 24). See Goldstein et al. (2011) for an excellent didactic overview and Maudlin (2011) for a discussion of the consequences of this theorem.

ment outcomes elsewhere in space. Bell's theorem is not limited to quantum mechanics: it puts a constraint on any physical theory that is to be empirically correct in yielding the experimentally confirmed predictions of quantum mechanics. No such theory can consider only the parameters that are situated in the past light cone of a given event to make correct predictions of the event in question and its future.

To be precise, Bell's theorem is derived from two premises: in the first place, there is the mentioned locality premise according to which all that enters into the probabilities for certain events to happen at certain space-time points is situated in the past light cones of these points. Furthermore, there is a premise to the effect that the past of the measured system is independent of the fixing of the parameters to be measured on the system. This premise is known as "no conspiracy". It has nothing to do with the free will of experimental physicists. The choice or fixing of the parameters to be measured can be carried out by a computer. All that is required is that this fixing is not correlated with the past of the measured system.

Also in a deterministic physical theory, this premise is satisfied. A deterministic theory only says that given the laws and initial conditions, the evolution of the systems under consideration is fixed. But determinism does not impose any restrictions on the initial conditions. This is done only in a position that is known as super-determinism. This is the claim that not only the evolution of physical systems is deterministic; furthermore, also the initial values of all the parameters in the past back to the initial state of the universe are correlated with each other. Consequently it is not possible to have a past state of a given system in the universe and fix variables to measure on that system in a way that is independent of its past state. Only super-determinism, but not determinism contradicts the "no conspiracy" premise.[34]

"No conspiracy" is a general premise that applies to any experiment. No experiment could give us information about the measured system if the system conspired with the questions that it gets asked in the guise of the choice of the parameters that are measured on it. There is nothing specific about the experiments in quantum mechanics in this premise. Hence, the only specific physical premise in Bell's theorem is locality.

---

[34] See Esfeld (2015).

This theorem and the subsequent experiments therefore establish that quantum physics violates locality.

This non-locality distinguishes quantum from classical theories, including classical electrodynamics and general relativity. It is the feature that one cannot avoid when passing from classical to quantum physics. One does not have to change the (primitive) ontology of point particles that are characterized by their relative positions only and whose configurations evolve according to a deterministic law of motion. One does not have to change the status of probabilities either. Refusing to make any such change precisely is the means to avoid the conceptual puzzles that are associated with quantum mechanics. Hence, one can retain the (primitive) ontology and the procedure to derive probabilities from deterministic laws, but one has to change the dynamical laws when passing from classical to quantum physics.[35]

While classical electrodynamics and general relativity are local field theories, Newtonian gravitation also is a non-local theory. However, quantum non-locality is in at least two important respects different from the non-locality that is implemented in Newtonian gravitation: (1) it is independent of distance in space, and (2) it is selective—it affects only those physical objects whose wave-function is entangled. As mentioned at the beginning of Sect. 1.5, the non-locality in Newtonian gravitation is often described as "action at a distance", whereas this term is usually not employed in the context of quantum non-locality. But this is a misunderstanding. In both cases, what happens at a particular location in space depends on what there is elsewhere in space at the same time without there being anything that is transmitted across space.

In the Newtonian case, the non-locality concerns all objects, but diminishes with the square of their distance. In the quantum case, the non-locality is independent of distance, but it is selective—it does not concern all physical objects indiscriminately, but only those whose wave-function is entangled. Such restrictions are indispensable for the theory to be in the position to explain why we usually are not aware of non-locality in common experience, neither in the case of Newtonian gravita-

---

[35] Furthermore, see Chen (2019) for a proposal how to integrate thermodynamics and the past hypothesis in quantum physics.

tion, nor in the quantum case. Usually, objects that are far away in space can be neglected for all practical purposes in Newtonian gravitation, as quantum entanglement can often be neglected for all practical purposes. Sophisticated experiments such as the ones confirming Bell's theorem have to be designed to make quantum non-locality manifest.

Again, Bohmian mechanics explains these facts: like Newtonian mechanics and like any other fundamental physical theory, it is a theory about the universe as a whole. That is to say, the particle configuration that it considers is the configuration of the universe, and the wave-function is the—entangled—wave-function of the universal particle configuration. But the theory then contains a precise mathematical procedure to get from the hypothesis of a universal wave-function to wave-functions for subsystems in the universe and to define circumstances under which these wave-functions are not entangled for all practical purposes; they then are known as effective wave-functions and can be calculated so that laboratory experiments can be prepared according to them.[36]

In general, Einstein (1948, pp. 321–322)[37] is right in pointing out that if non-locality were ubiquitous, physics in the sense as we know it would not be possible; for it would then be impossible to isolate physical systems and make experiments only on them. However, contrary to what Einstein maintained, this is no argument against quantum physics; for the non-locality in the quantum case is limited, as it is already limited in the Newtonian case. Nevertheless, Einstein had every reason to be concerned about quantum non-locality: interactions in quantum theory do not comply with the principle of local action according to which everything that can influence what happens at a space-time point is situated in the past light cone of that point. This principle is implemented in classical electrodynamics. Einstein then showed with the general theory of relativity how a theory of gravitation can be set out that satisfies this principle and supersedes Newton's theory. As is proven by Bell's theorem, in quantum physics, by contrast, what happens at a space-time point can also be influenced by what happens outside the past light cone of the event in question. Bell (2004, ch. 24) and the seminal monograph by

---

[36] See Dürr et al. (2013, chs. 2 and 5).
[37] English translation in Howard (1985, pp. 187–188).

Maudlin (2011) elaborate on the conflict between quantum non-locality and relativity physics.

That notwithstanding, quantum field theory is supposed to unify quantum mechanics with special relativity theory. For instance, quantum electrodynamics is a relativistic quantum theory of electrodynamics. Quantum field theory is supposed to be refoliation invariant: it does not rely on a privileged foliation of four-dimensional space-time into three-dimensional spatial hypersurfaces. However, this is true only as far as the algorithm for the calculation of measurement outcome statistics is concerned. When one asks what happens in nature—that is, asks for a theory that accounts for the occurrence of individual, determinate measurement outcomes—, as things stand, any such theory is committed to a privileged foliation of space-time. This is so independently of whether or not the theory assumes the collapse of the wave-function in its dynamics.[38] The reason for this situation is the non-locality proven by Bell's theorem and confirmed by numerous experiments. It is only that this privileged foliation is not accessible in experiments because it is not possible to predict individual measurement outcomes due to the principled lack of exact knowledge of initial conditions.

One may hence employ quantum non-locality as an argument to resist the conclusion from special and general relativity to the ontology of a block universe. However, drawing on quantum physics is not the best argument to resist that conclusion, since it brings in considerations from another area of physics. The better physical argument is the one discussed at the end of Sect. 1.6: one can trade refoliation invariance for scale invariance. That is to say, one can obtain a theory of gravitation that yields the same empirical results as general relativity theory on the basis of working with instantaneous spatial configurations and thus an absolute distinction between space and time, but only employing relational and thus scale invariant quantities, as done in shape dynamics. By the same token, one can embed a quantum theory with an ontology and dynamics for individual physical systems, such as Bohmian mechanics, in shape dynamics, that is, by working only with relational and thus scale invariant quantities, instead of conceiving the particle configuration as

---

[38] See Barrett (2014) and Esfeld and Gisin (2014).

being inserted in an absolute space and time.[39] That notwithstanding, it is an ongoing research project whether shape dynamics can provide a dynamics for the whole of physics (including a future theory of quantum gravity).

As regards the ontology of quantum field theory, as in quantum mechanics, it would be naïve to seek to read ontological conclusions off from the textbook formalism to calculate measurement outcome statistics, on pain of being trapped in the conceptual puzzles, such as the measurement problem.[40] In particular, quantum field theory does not commit us to an ontology of fields. It does not work with fields in its dynamical structure as in the classical sense of determinate values that are attributed to space-time points. The fields in quantum field theory are operator valued fields that are employed as instrument to calculate measurement outcome statistics. In general, quantum *field* theory is the dynamical framework for today's standard model of elementary *particle* physics.

As in quantum mechanics, one can embed the formalism of quantum field theory into a primitive ontology of a fixed number of permanent point particles that move on trajectories according to a deterministic law of motion; modulo a probability measure in the guise of quantum equilibrium, one can then derive from this ontology and law the quantum field theory textbook formalism to calculate measurement outcome statistics (which, in this case, works with a varying number of appearing and disappearing particles, detected in scattering experiments).[41] Again, the argument for doing so is that one thus solves the measurement problem also in the domain of quantum field theory. In brief, at the end of the day, there is no cogent reason in quantum physics—neither in quantum mechanics, nor in quantum field theory—to abandon the primitive ontology of configurations of point particles and their change that is at the bottom of the success of modern physics.[42]

---

[39] See Dürr et al. (2018) for a detailed treatment. See furthermore Vassallo (2015), Vassallo and Ip (2016) and Koslowski (2017).

[40] See Barrett (2014) for the measurement problem in quantum field theory.

[41] See Esfeld and Deckert (2017, ch. 4) for the details how to do this. See also already Colin and Struyve (2007).

[42] Cf. also the quotation from Feynman in Sect. 1.1.

Indeed, the success not only of modern physics, but of modern science as a whole relies on atomism. In this chapter, I have conceived atomism as a fundamental or primitive ontology that is characterized only by the following two axioms or principles:

1. *There are distance relations that individuate simple objects, namely point particles (matter points).*
2. *The point particles are permanent, with the distances between them changing.*

These two axioms or principles define an ontology of the natural world that answers the following question: what is an ontology that is minimally sufficient to accommodate what science as well as our common sense knowledge tell us about the natural world? The considerations about classical mechanics and electrodynamics, about relativistic physics and about quantum physics in this chapter are intended to show that there is indeed an answer to this question that captures the whole of modern physics and that is expressed by these two axioms or principles. If one were to enlarge the ontological commitments, one would not only fail to achieve a deeper understanding of nature and natural science, but also run into new problems that are pseudo-problems, to use again the term of Carnap (1928). To be clear about the fact that this minimalist ontology is sufficient to understand modern science will turn out to be the key for realizing in the following chapters that modern science paves the way for our freedom instead of infringing upon our free will.

Everything else enters a physical theory as part of the dynamical structure that is needed to formulate laws of motion for the material objects and to make predictions, and be it statistical predictions by means of a probability measure given our ignorance of the exact initial conditions. The dynamical structure does not call for additional ontological commitments. It is not the guide for ontology. Simplicity and richness in information in the description of the motion of matter do not go together with simplicity and clarity in ontology, but are opposed to the latter. The reason is that in order for a description to be simple and informative about the motion of matter, much more means of representation are needed than those ones that are minimally sufficient to understand what matter is and what it does.

This claim will be vindicated in the next chapter. The following chapter will work out how scientific explanations are structured, what they achieve and what they do not achieve and what is the—epistemological as well as ontological—status of laws of nature. This will lead at the end of the next chapter to the argument why science is not opposed to our free will, but, quite to the contrary, enhances our freedom.

2

How Science Explains: Scientific Explanations and Their Limits

2.1 The Location Problem and Its Solution: Functionalism

Consider how the contemporary Australian philosopher Frank Jackson describes the task of philosophy, viz. ontology or metaphysics:

> Metaphysics, we said, is about what there is and what it is like. But of course it is concerned not with any old shopping list of what there is and what it is like. Metaphysicians seek a comprehensive account of some subject matter—the mind, the semantic, or, most ambitiously, everything—in terms of a limited number of more or less basic notions. In doing this they are following the good example of physicists. The methodology is not that of letting a thousand flowers bloom but rather that of making do with as meagre a diet as possible. ... But if metaphysics seeks comprehension in terms of limited ingredients, it is continually going to be faced with the problem of location. Because the ingredients *are* limited, some putative features of the world are not going to appear explicitly in the story. The question then will be whether they, nevertheless, figure implicitly in the story. Serious metaphysics is simultaneously discriminatory and putatively complete, and the combination of these two facts means that there is

© The Author(s) 2020
M. Esfeld, *Science and Human Freedom*,
https://doi.org/10.1007/978-3-030-37771-7_2

bound to be a whole range of putative features of our world up for either elimination or location.[1]

This is exactly what atomism does, namely to provide a comprehensive account of the natural world "in terms of a limited number of more or less basic notions". Everything in the natural world is composed of spatially arranged point particles and can be understood on the basis of the interactions of these point particles, that is, how their spatial arrangement changes in time. To use the famous term of Sellars (1962), this is the *scientific image of the world*.

If one spells atomism and this image out according to the methodology of "making do with as meagre a diet as possible", one gets down to the two axioms mentioned at the end of Sect. 1.2, which define the basic or primitive ontology:

1. *There are distance relations that individuate simple objects, namely point particles (matter points).*
2. *The point particles are permanent, with the distances between them changing.*

The methodology of "making do with as meagre a diet as possible" should not be received as being concerned with necessity and *a priori* knowledge. The point at issue is minimal sufficiency. The task is to answer the following question: What is an ontology that is minimally sufficient to accommodate what science as well as our common sense knowledge tell us about the natural world? To this day, the only worked out answer to this question is atomism whose essentials are captured by these two axioms. Nevertheless, there may be other answers possible. If so, we would face a problem of underdetermination of the ontology of science—but only when alternative answers were worked out in detail.

In particular if the ontological diet is as meagre as possible, the problem is evident how to find everything that science and our common sense knowledge tell us about in this ontology. As Jackson makes clear, this problem of location, which is also known as placement problem (Price

---

[1] Jackson (1994, p. 25). See also Jackson (1998, ch. 1).

2004), arises in any case if one formulates a scientific or philosophical theory of the world. Abandoning the methodology of parsimony and admitting more things as primitive than what is minimally sufficient would be of no help with that problem. The task is to formulate a general recipe how to locate or place something that does not figure explicitly in the basic notions that define the ontology, however meagre or abundant these notions may be. If one has obtained a solution to this problem, it will apply to whatever comes up for location.

The solution to the problem of placement or location is provided by the stance that is known as functionalism in the philosophy of science. Again, this may be not the only possible answer, but it is the only fully worked out one. Starting from configurations of point particles as described by the basic notions that fix the primitive ontology, one defines everything else in terms of its function in the sense of its role for the evolution of point particle configurations; this then enables the location of the thus defined things in configurations of point particles, namely in those ones that realize the role in question, as is clear since the 1970s, when notably David Lewis (1970, 1972) set this method out.

Consider water. As we know from scientific investigation, there is no fundamental water stuff in the world. Science superseded the ancient view of the four elements earth, water, air and fire. But, of course, there is water in the world: there are things in the world that fulfil the functional role of appearing odourless, colourless, being thirst-quenching through the change in the motion of the parts of our bodies that they bring about. These are configurations of H_2O molecules.[2] Thus, by defining water in terms of its thirst-quenching role—that is, its role for certain motions in our bodies—, we locate water in the primitive scientific ontology of atomism. Certain particle configurations, moving in certain characteristic ways, *are* water.

By the same token, there is no *élan vital*, a *sui generis* life stuff or causal power; but there are organisms in the world. The functional role that defines what it is to be alive in terms of characteristic motions such as

---

[2] For the sake of the example, let us leave aside here the view according to which "water" is a rigid designator such that the water-role can be realized only by H_2O molecules. See Putnam (1975) for this view.

reproduction and adaptation to the environment is realized by certain configurations of molecules, as we know since the rise of molecular biology in the twentieth century. Consider as one famous example the discovery of the molecular composition of the DNA by James Watson and Francis Crick (1953). Again, this means that certain particle configurations, moving in certain particular ways, *are* organisms. Life thus is located in certain particle configurations. There is no additional, primitive life stuff.

Furthermore, there arguably are no *sui generis* minds; but there are mental states defined by certain functional roles, which in the end are functional roles for the behaviour and thus the bodily motions of persons, realized by certain neuronal configurations in the brain. This is the view of functionalism, which is the mainstream stance in today's philosophy of mind and cognitive science and is clearly brought out by Jaegwon Kim (1998) among others. Thus, the driving idea behind much of today's research in this field is that neuroscience will do for the mind what molecular biology did for life, namely to open up the mind for scientific investigation by discerning the neuronal configurations that realize the functional roles that define mental states. Again, this means that certain particle configurations—in this case, certain neuronal configurations—, moving in certain particular ways, *are* minds. In short, minds then are located in brains so that there are no additional, primitive minds over and above matter.

Functionalism does not only apply to the objects of the special sciences, that is, all sciences apart from the fundamental and universal theories of physics. If the ontology is given just by the two above mentioned axioms, then all there is to the point particles are the distance relations in which they stand and their change. Hence, already everything that enters into the dynamical structure of a physical theory over and above the parameters that designate the primitive ontology does not count among the basic notions. Instead, it is introduced in a functionalist manner in terms of its causal role for the elements of the primitive ontology. Consequently, the task then is already to locate for instance mass in the configurations of point particles individuated by distance relations and their change.

Consider gravitation: the motion of the objects in the world manifests some salient patterns or regularities. Arguably the most striking of these patterns is mutual attraction. As Feynman et al. (1963, ch. 1–2) put it in the quotation at the beginning of Sect. 1.1, "*all things are made of atoms— little particles that move around in perpetual motion, attracting each other when they are a little distance apart, but repelling upon being squeezed into one another*". This pattern of motion applies everywhere and at every scale in the universe, from atoms to apples falling from trees, stones falling to the soil on the Earth, to planetary motion, such as the motion of the Earth around the Sun. This stable pattern enables us to introduce the notion of gravitational mass in order to represent this regular motion: gravitational mass is defined in terms of its function for particle motion, namely the function of mutual attraction. As already quoted in Sect. 1.3, Ernst Mach brings this functional introduction of the notion of mass out in his comment on Newton's *Principles* when saying "The true definition of mass can be deduced only from the dynamical relations of bodies" (Mach 1919, p. 241).

All the evidence that we have are the dynamical relations of bodies— that is, their motions; these relations manifest certain stable patterns, such as attractive motion. To represent this pattern in a theory, physicists introduce the parameter of mass as defined by its function for the particle motion. Function means here functional or causal role, namely the role for the change in the particle configuration as defined by the spatial arrangement of the particles. Having such a parameter at one's disposal then enables the formulation of a law that captures the pattern at issue, such as Newton's law of gravitation. By means of the parameter of mass, as well as a constant (namely the gravitational constant), this law makes it possible to describe, calculate and predict the attractive motion of bodies.

There are more stable patterns in the motion of bodies than gravitational attraction. There is a further characteristic pattern of repulsive and attractive motion that also applies at all scales, namely the pattern of electricity and magnetism. To represent this pattern in a theory, one introduces a further parameter that is defined by its function for the particle motion, namely the parameter of charge. By means of this parameter, one can then formulate the laws that make it possible to describe, calculate and predict this characteristic repulsive and attractive motion of bodies (i.e. the Lorentz force law and the Maxwell equations in classical electrodynamics).

By means of this procedure of a functional definition of parameters such as mass and charge, one locates mass and charge in what is accepted as primitive, namely the particle motion, through the fact that the particle motion manifests certain stable patterns or regularities. Given the fact of such salient patterns or regularities, there is no need to include parameters such as mass and charge among the ontological primitives and to consider mass and charge as intrinsic properties of the objects—that is, as something that the objects possess in and of themselves. They enter a physical theory through their role for the particle motion.

Nevertheless, they are thereby admitted to the ontology, albeit not as primitive, but as derived notions. The particles have mass and charge not as primitive features, but because they move in certain manners. In virtue of their motion particles have mass and charge, in the same way as some particle configurations are water, organisms or minds in virtue of their motion (if functionalism is true about the mind as well). All these features of the world are literally located in the particle motion.

Four comments about the functionalist solution to the problem of location are noteworthy:

1. Locating these features in certain particle configurations so that they are nothing over and above these particle configurations does not mean that these features are something that is intrinsic to these configurations. If one endorses functionalism with respect to the physical parameters that are not part of the primitive parameters of distances and their change, the propositions that ascribe certain values of mass and charge to individual particles are not true in virtue of these being intrinsic features of the particles, but in virtue of certain stable regularities in the overall particle motion. By the same token—though less obviously so—, no particle configurations are intrinsically water, organisms or minds. Certain particle configurations can be water, organisms or minds only if they are inserted in an environment with certain stable conditions—that is, certain stable regularities—such that these configurations can exercise the functional roles that define

water, organisms or minds. The environment is, strictly speaking, the entire rest of the universe: this condition defines normal conditions for the exercise of these functional roles in terms of nothing from the rest of the universe preventing the stable regularities in question from obtaining, such as the regularity that leads from H_2O molecules to thirst quenching motions in the body, or from certain chains of molecules to motions that are phenotypic traits of certain organisms, etc. Hence, to put it in a nutshell, locating features of the universe that do not explicitly figure in the notions defining the primitive ontology always is a holistic affair, although these features are located in certain particle configurations.

2. The idea of the functionalist solution to the problem of location is to define everything that does not figure explicitly in the basic notions in terms of its function in the sense of its causal role for the development of the primitive ontology as defined by the basic notions. In a sense, this is just a matter of definition. One can simply stipulate that everything else be defined in terms of such a causal role. However, this cannot be any odd causal role, but the definition has to point out a causal role for the development (evolution, behaviour) of what is admitted in the primitive ontology, that is a causal role for motion that finally is particle motion. Otherwise, the problem of location would not be solved. There then can be a meaningful debate about whether or not a functional definition in this sense captures the features of the world that it targets. There is such an ongoing debate about the mind. We will consider it in Chap. 3.

3. It cannot be functional roles all the way down. Functional roles always are roles for the behaviour of something that realizes the role in question. If all the realizers were themselves functionally defined, the position would run into a *reductio ad absurdum* in the guise of an infinite regress of realizers. By the same token, it could not be information all the way down, since information always is information about something. Thus, laws of nature, for instance, are informative, because they indicate constraints on the possible motions of matter; but the motion of matter itself is not information. Likewise, it could not be structures all the way down, since structures are relations that require relata as that what stands in the relations; if one takes all these relata to be

dissolved into further relations, one runs into an infinite regress, as the moderate ontic structural realism advocated by Esfeld (2004) and Esfeld and Lam (2008) argues against the radical ontic structural realism of French and Ladyman (2003). That is why functionalism cannot be a basic or starting position. The starting point for any serious philosophical project always has to be the postulation of a primitive ontology, namely a stance about what simply exists in the world (or the domain of the world under investigation), described by the basic notions. Functionalism then comes in as the solution to the problem of how to integrate into this ontology the features of the world that do not figure explicitly in it.

4. Employing the term "causal role" in this context does not commit one to a particular view about causation. This term is used here merely to demarcate functions in the sense that is relevant here from functions in the mathematical sense. The sense at issue here is role for the change (behaviour, evolution, development) of something. The term "causal" only is intended to emphasize that the issue is such a role.

This, then, is the *scientific image* of the world. The method of science is to postulate basic entities—such as particles in motion—and to explain everything else on that basis, namely as being composed of these particles and as consisting in a certain functional role for in the last resort the motion of these particles. Thus, for instance, a table or a cat is composed of particles. However, an ephemeral table-shaped or cat-shaped particle configuration is not a table or a cat. For a table-shaped or a cat-shaped particle configuration to be a table or a cat, it has to behave—that is, to move—in the way that is characteristic of a table or a cat. Thus, the crucial point for something to be a table or a cat is to fulfil the functional definition of what it is to be a table or a cat. This, again, cannot be an intrinsic feature of certain particle configurations, but depends on the conditions in the environment of the configurations in question. As illustrated by these examples and the ones given above, this procedure applies to everything apart from the basic entities that define the primitive ontology.

The problem of location and its functionalist solution do not only concern the special sciences, such as chemistry (example water), biology (example organisms) and cognitive science (example mind). This problem occurs and this solution applies already in physics: parameters such as mass and charge are introduced in terms of their functional role for the particle motion. Consequently, the different particle species distinguished in today's standard model of elementary particles do not indicate intrinsic features of the particles. They depend on the way in which the particles move under given, stable environmental conditions that typically obtain in the universe. In brief, some particles are electrons, because they move electronwise so to speak under standard conditions.

Nonetheless, for the working physicist, it is much easier to formulate the physical theory by labelling the particles "electrons", "quarks", etc. That notwithstanding, upon reflection, it becomes evident that the parameters that allow for the differentiation into several particle species (such as mass, charge, spin) are not primitive, but enter the physical theory through their functional role for the particle motion. This, consequently, then is the motion of naked particles so to speak—that is, particles that do not have any intrinsic features, but all there is to them are their relative positions and the change in these positions. In particular, in quantum physics, also classical parameters such as mass and charge are situated on the level of the wave-function.[3] Due to entanglement, the wave-function cannot be applied to the particles taken individually.

Again, this can be nicely illustrated in terms of the Bohmian quantum theory. Strictly speaking, the correct theory is the one known as identity-based Bohmian mechanics, which treats all particles as identical in its formalism in the sense that they are not differentiated into several species from the outset in virtue of their mass, charge and spin. But the formalism of this theory is much more complicated to handle than the one of the standard Bohmian mechanics. Both theories are empirically equivalent as they lead to Born's rule for the prediction of measurement outcome statistics. But they differ in the trajectories that they attribute to the particles. In order to understand why there is no arbitrariness in the notion of Bohmian particle trajectories, it is then important to go down to the version that follows from the ontology of the theory. This is the

---

[3] See Brown et al. (1995, 1996) and later Pylkkänen et al. (2015) as well as Esfeld et al. (2017).

identity-based Bohmian mechanics, which sets out a formalism that treats the particles as being characterized by their relative positions only.[4]

By the same token, for the scientist in the special sciences, it makes handling the theories much easier to pretend that there are intrinsic chemical elements, intrinsic biological species, etc. That this is not so, that these elements or species do not have to be accepted as something primitive, but are functionally defined in terms of the roles of their characteristic features for in the last resort particle motion becomes relevant only when it comes to the relationship between the special sciences and physics. That is, it becomes relevant only when one reflects upon the unity of science. This is both an ontological unity (identity with particle configurations) and a methodological unity (functional definitions).

Science is automatically committed to a reductionism in the following sense: any theory has to locate features of the world—or the domain of the world under investigation—that do not figure explicitly in the notions that are endorsed as basic; that means tracing these features back or reducing them to what is described by the basic notions. The solution to this problem consists in *identifying* some configurations of the entities that are admitted as basic with these features. Thus, some particle configurations *are* water, others *are* organisms, etc., given certain normal background conditions in the environment. Identity is symmetrical: if certain particle configurations are identical with the water that there is in the universe by playing the water role against normal background conditions, then the water that there is in the universe is identical with certain particular particle configurations. Despite being symmetrical, this identity amounts to an ontological reduction: everything is particles and their configurations (that is, reduced to particles and their configurations), whereas only some specific particle configurations are water, organisms, etc.

This ontological reductionism goes with an epistemological reductionism. Once the functional definitions of water, organisms, etc. in terms of in the last resort their roles for particle motion are given, it then follows from a complete description of the particle motion which particle configurations are water, which ones are organisms, etc. Jackson (1994, 1998, ch. 3)—preceded by Lewis (1966, 1970, 1972)—speaks even of *a*

---

[4] See Goldstein et al. (2005a, b) and Esfeld et al. (2017).

priori entailment, since he takes the functional definitions to be a matter of conceptual analysis, which is an *a priori* affair. Thus, a complete description of the universe in terms of the basic physical notions entails all the true statements about the world, given functional definitions of all the other features of the world. However, this is only an in principle epistemological reductionism, which is nothing more than a logical consequence of the solution to the location problem. It cannot be transformed into a recipe how to reduce all our scientific knowledge to fundamental physics.

No one has—and will never have—a complete description of the basic physical entities of the universe at one's disposal. The features of the universe that the special sciences capture are salient features or patterns existing on Earth with there being no physical concepts that bring out how these features are salient. Physics has no particular interest in precisely those particle configurations that are water, organisms, etc. That is why the special sciences are indispensable not only for all practical purposes, but also for cognitive reasons. To put it in a nutshell, the reductionism that comes with the methodology of science is conservative by contrast to eliminative, as argued by Esfeld and Sachse (2011): the issue is not to eliminate features of the world or concepts of the theories of the special sciences, but to locate them in what is accepted as basic.

It is trivial that there are new features coming up in the evolution of the universe, that is, features which are limited to specific places and times, such as the formation of water molecules, or the development of organisms, etc. But these are explained in terms of the dynamical laws that apply everywhere in the universe (as far as we know) plus special initial conditions, which, again, are special initial conditions of the universe in the last resort. Recall the past hypothesis, which states that the initial particle configuration of the universe is one that implements a very low entropy (see the discussion in Sect. 1.4). Hence, these are not new ontological features of the universe: by means of such an explanation, they are located in the particle configuration and its evolution (provided that they admit of a functional definition). In brief, far from being opposed to reduction, emergent features in the sense of new features coming up in the evolution of the universe just are a central subject matter to which the methodology of location applies.

One can employ the notion of emergence also in a philosophical sense that is opposed to reduction. In this case, one is committed to new features that count as further primitives in the ontology. If there are features of the world that do not admit of location in what is accepted as primitive, there are only two options, as comes already out in the quotation from Jackson (1994) at the beginning of this chapter: either one eliminates these features—that is, gives an argument why it is an illusion to believe that they exist—, or one endorses these features as further primitives, over and above matter in motion. The logic is this one: for anything, the thing in question either exists, or it does not exist. If it exists, it either belongs to the primitive ontology, or it is derived from the primitive ontology (in the sense of being identical with certain configurations of elements of the primitive ontology). Hence, if one is committed to the existence of something without being able to derive it from the primitive ontology, one has to enlarge the primitive ontology so that it includes this thing or feature as a further primitive. The notion of emergence then is of no help to understand such features. It is irrelevant here whether such features occur only at certain places or times. If they exist and cannot be located in what is accepted as primitive, there is no other possibility but to endorse them as further primitives in the ontology, however rare or abundant their occurrence in the universe may be.

We obviously face that debate when it comes to the mind, and we will go into it in the next chapter. There is no point in seeking to avoid that debate by employing a confused notion of emergence—that is, a notion that takes emergence to be opposed to reduction, but bases itself on the trivial sense of the emergence of new features at specific places and times in the universe. The confusion then lies in the suggestion that there can be the emergence of something within the framework of the scientific image of the world without that something being located in the ontology and seized by the methodology on which this image is based. However, being clear about the scientific image of the world and what it implies is the presupposition for conducting a meaningful debate about whether there exists something that is not accounted for in the scientific image and how, in case this is so, one can recognize these things while at the same time recognizing the truth about the world that is discovered within the scientific image.

2.2   What Scientific Explanations Achieve and What Their Limits Are

The functionalist solution to the problem of location provides an explanation of everything to which it applies. For instance, if one asks why there is water in the universe, the answer is that there are H_2O molecules bound together in a specific manner and behaving under standard environmental conditions in such a way that they fulfil the causal role that defines water. This answer applies to the general question why there is water at all in the universe ("Why is there water?") as well as to the specific question why a given specimen is a sample of water ("Why is this water?"). This answer is a causal explanation: it shows how particle configurations of a certain type realize under standard environmental conditions a certain causal role, namely the causal role that defines a certain specific feature of the universe (such as the feature of being water). An explanation of this type applies to all the features of the universe that the special sciences treat, insofar as these features enter into the scientific image of the world. More precisely, these features figure in the scientific image of the world because they come under an explanation of this type.

Again, this is not to say that such an explanation can be carried out in practice in each single case. However, if one has reasons to presume that a certain feature of the world does not come under an explanation of this type, these are necessary and sufficient reasons to reject a functional definition of the feature in question in terms of a causal role for the motion of some matter. By way of consequence, the feature in question then is not part of the scientific image of the world and calls for an ontological commitment to further primitives beyond what is admitted as primitive in the scientific image.

The debate about consciousness can again serve as a good example. If one maintains that consciousness is characterized by intrinsic, qualitative features (so called qualia), then one rejects a functionalist definition of consciousness. One thereby excludes that consciousness can be fully treated by scientific methods and is part of the scientific image of the world. By the same token, functionalism is the mainstream in contemporary cognitive science, because it promises to apply scientific methods to consciousness and thereby to establish that consciousness (and the mind as a whole) is part and parcel of the scientific image.

Functionalist explanations are not limited to the special sciences. They apply already in physics. Also universal dynamical parameters such as mass and charge are introduced in physics through their function in the sense of their causal role for the motion of matter. Hence, if one asks why there are objects with mass or charge in the world, the answer is that there are objects that move in such a way that their patterns of motion realize the causal roles that define mass or charge.

Nonetheless, one may have the impression that something has gone wrong now, namely that the usual scheme of explanations has been turned upside down. Why, for instance, does the Earth move around the Sun? The usual answer is this one: both the Sun and the Earth have mass. The mass of the Earth is much smaller than the mass of the Sun. In virtue of this ratio between the masses, the Earth is attracted by the Sun. This looks like a causal explanation such that the mass brings about the attractive particle motion.

Following the functionalist definition of dynamical physical parameters such as mass and charge, by contrast, one has to say this: there are the particle motions as described by the basic physical vocabulary, that is, the notions defining the primitive ontology of relative distances among the particles and their change. Certain of these particle motions described in that manner fulfil the causal roles that functionally define features of the universe that do not figure explicitly in the description of the universe by the basic physical notions. Hence, this explanation endorses the particle motion as described in the basic physical vocabulary as primitive. On this basis, it explains everything else in causal terms in the sense of certain of these particle motions realizing specific causal roles as described in the non-basic, functional vocabulary. This, then, is to say that the attraction of the Earth by the Sun being part of the general pattern of attractive motion explains why the Earth and the Sun have certain values of mass respectively. By the same token, the characteristic motion of certain particles making up H_2O molecules explains why these molecules are water. The same goes for all the features of the world that are located in the primitive ontology of physics through functional definitions.

By way of consequence, causal explanations end once we have reached the basic physical patterns or regularities of motion, which apply all over the universe as far as we can tell. These stable patterns or regularities are the bedrock. There is no causal explanation within science available why they obtain. That notwithstanding, there still are explanations at this basic level. But these are not explanations of why there are certain types of motion such as attractive motion at all, but only of particular tokens. Thus, the explanation of why the Earth moves around the Sun is that this motion is an instance of a general pattern of attractive motion that holds everywhere in the universe, for instance also in the apples falling from the trees. This is explanation in terms of unification: one shows why a particular phenomenon is not astonishing because it comes under a general pattern.[5]

Upon reflection, it is evident that explanations in science have to end somewhere and that they end when the basic regularities that obtain in the universe are reached. We observe certain regularities in the natural world, such as the regularities of attractive motion of bodies summed up under the term "gravitation", or the regularities of attractive and repulsive motion summed up under the term "magnetism". We trace these observed regularities back to the basic entities that compose the observed bodies (that is, the point particles) and their motion as described by the basic physical notions. For there to be stable particle configurations in the universe some of which then are molecules and finally organisms up to human beings, there have to be certain stable patterns or regularities in the way in which the particles move. Since these are patterns in the particle motion, they are described by the basic physical notions admitted as primitive. This then implies that there is no further explanation of these regularities available within the scientific image of the world, apart from the mentioned explanation by unification (that is, showing how particular phenomena are an instance of these universal regularities).

It would be futile to search for further explanations within the scientific image of the world. The reason again is this one: apart from the basic physical notions defining the primitive ontology, all the other notions are

---

[5] See Friedman (1974) and Kitcher (1989). See furthermore Bhogal (2019, Sect. 2.1) for the link between explanation by unification and the metaphysics of laws.

introduced through the outlined procedure of functionalization, namely through their causal role for what is described in terms of the basic notions. More precisely, anything that cannot be introduced in this way is part of the basic notions, whereas everything that can be introduced in this way is thereby located in what is described in terms of the basic notions. That is the reason why already universal physical parameters such as mass and charge are not part of the basic vocabulary and do not belong to the primitive ontology: they can be defined functionally in terms of their role for the particle motion. By contrast, notions such as "distance" and "change" admit of no functional definition in terms of their role for something more basic. That is why they define the primitive ontology of the scientific image of the world.

Notions that are introduced by this functionalist procedure cannot answer the question why there is the basic change for which they perform a certain role, on pain of circularity. The circularity is the one at which Molière pokes fun in his piece *Le malade imaginaire*: one does not explain why people fall asleep after the consumption of opium by attributing a dormitive power to opium, because this power is *defined* in terms of the role to make people fall asleep. By the same token, one does not explain why there is attractive motion in the universe by attributing a mass to bodies, because mass is *defined* in terms of the role to make bodies attract each other.

Of course, mass and charge are sparse, universal physical parameters by contrast to phenomenological properties of opium. Nonetheless, Molière's argument hits also these physical parameters, if they are conceived as powers: like the dormitive power of opium, they are defined in terms of the effects that they bring about under certain conditions. That is why they cannot explain these effects. And that is why it does not lead to a gain in explanation to admit powers or dispositions as ontological primitives.[6] They cannot explain why there is motion at all, for they are powers for specific motion. They cannot explain that specific motion either, for they are defined in terms of a causal or functional role that is precisely that specific motion. Furthermore, the commitment to parameters such

---

[6] See Bird (2007) and Mumford and Anjum (2011) as well as the papers in Marmodoro (2010) for prominent such positions.

as mass and charge being properties of objects in the guise of dispositions or powers again implies a commitment to surplus structure: these properties can be there without manifesting themselves in a making an empirical difference, for instance in a situation in which mass and charge cancel each other out so that there is no acceleration of objects.

Endorsing powers or dispositions as ontological primitives not only fails to improve on explanations, but also leads to pseudo-problems that have no solution. Suppose that mass and charge are intrinsic properties of particles that are the powers to bring about the motion in terms of which they are defined. Then, how does a power that is intrinsic to a particle reach out of that particle and change the motion of *other* bodies at a distance? Invoking fields as mediators is of no help in answering this question, since it leads to the physical problem of self-interaction and to the philosophical problem of the unclear ontological status of fields, as set out in Sect. 1.5.

The dynamical laws, in which parameters such as mass and charge figure, are connected with the symmetries of space-time. However, admitting these symmetries as ontological primitives in the guise of ontic structures faces again the same problem: they are defined in terms of the constraints that they impose on the motion of the bodies in the universe. Hence, posing these symmetries or structures as ontological primitives does not explain why there are such constraints on the motions of bodies, that is, why the motion of bodies manifests regularities that satisfy these constraints.

There is no question here of a limitation to normal, or ideal conditions such that there could be a discrepancy between what the objects *de facto* do and what they would do if the conditions were normal, or ideal. If one conceives mass and charge as powers, they are powers that are attributed to the objects throughout the whole universe. By the same token, the symmetries or structures apply to the universe as a whole. They are defined in terms of the constraints that they impose, applying to the universe as a whole, which is the normal condition for their attribution. Hence, again, they come out as being defined by what they do in fact for the motion of the objects in the universe.

By contrast to admitting dynamical parameters such as mass and charge as dispositions or powers to the primitive ontology, one may

consider attributing a general, overall power for change to the configuration of matter in order to make the fact that there is change at all intelligible. For instance, Leibniz does so in the *Specimen dynamicum* (part I, §
1): he rejects the idea of continuous motion, endorsing what is today known as the at-at theory of motion (for which Russell (1914) is the central source in contemporary philosophy). That theory says that when an object moves, it is first here and then there, without moving continuously from here to there. In the framework of relationalism and applied to the configuration of matter of the whole universe, this is to say that a sequence or stack of spatial arrangements of matter of the universe occurs with a unique order of that sequence, but that sequence is not continuous. The overall power for change that is inherent in the configuration of matter then is what unites that sequence (given that there is no external, absolute time), accounting for the evolution from one arrangement to the next one. That power only unites that sequence. It does not amount to a deeper explanation of the particular change that occurs. In contrast to this view, the primitive ontology endorsed in this book admits continuous change in the configuration of matter as a primitive (axiom 2 in Sect. 1.2). It does therefore not need the commitment to such an overall power as that what unites the evolution of the configuration of matter. That evolution is united by being continuous.

An objection that is similar to the one against the conception of dynamical parameters such as mass and charge as dispositions or powers applies if one admits over and above the distance relations a space-time substance in which these relations and their change are embedded. In addition to the point particles (matter points), there then are space-time points, which are dense, forming the continuum of space-time as an ontological primitive. For instance, Maudlin (2007, pp. 87–89) takes the length of a path in absolute space as the primitive notion. He derives the notion of distance of point particles from that notion as the minimal path length connecting them, claiming that he is thereby able to explain the constraints on the distance relation (such as the triangle inequality, see Sect. 1.2). However, in order to be able to define a minimal path length in space, one has to presuppose a structure that is rich enough to accommodate a metric—as the relationalist has to presuppose a relation that is rich enough to fulfil the triangle inequality among others in order to count as distance relation. In short, substantival space comes with a metric in terms of, for example, paths of geodesic motion. Any metric defining a physical space is such that it fulfils all the constraints of three-dimensional geometry. Hence, there is no explanatory gain here in comparison to the relationalist who presupposes that the relations admitted as primitive fulfil the requirements that are sufficient for these relations to count as distances. There only is the disadvantage that substantival space contains more structure than is needed to accommodate the empirical evidence of relative positions and motions of bodies (see Sect. 1.3).

In sum, the space-time in which the configuration of matter is represented as being embedded, fields on that space-time (such as the electromagnetic field) and dynamical parameters (such as mass, charge, spin, a wave-function, etc.) are all part and parcel of the dynamical structure of a physical theory in distinction to the primitive ontology. They all come in through their role in the representation of the evolution of the elements of the primitive ontology—that is, the evolution of the distance relations in the particle configuration of the universe. They all are the means to achieve a representation of that evolution in terms of laws of nature that are both simple and informative. In a nutshell, they all come in as a package together with the laws.

Nonetheless, there is a notable difference between the physical space-time and fields on that space-time on the one hand and the dynamical parameters on the other hand. The described method of location in the particle configuration through a functional definition applies only to the dynamical parameters. As some particle configurations are water, organisms, etc. because they fulfil the functional definitions of water, organisms, etc., so particles have mass, charge, spin, etc. and are electrons, protons, neutrons, etc. in virtue of the way in which they move. Thus, having mass, charge, spin, being an electron, a proton, a neutron is located in the particle motion, namely in the salient patterns that the particle motion manifests. The same applies to constants of nature such as the constant of gravitation, the speed of light, etc.

The same goes also for the wave-function, which is the central dynamical parameter in quantum physics: as the particle motion makes it that particles are electrons, protons, neutrons and have a determinate mass, charge, spin, so the particle motion makes it that a determinate universal wave-function applies to the particle configuration of the universe (and derived from it effective wave-functions to subsystems of the universe under certain circumstances). Although the wave-function itself develops in time instead of being stationary, it and its evolution can be taken to derive from the particle motion in the same way as the mass that is attributed to the particles taken individually in classical physics derives from the way in which they move. In that sense, then, the wave-function is located in the particle configuration.[7] The field on configuration space only is its mathematical representation. This view is a realism about the wave-function by contrast to an instrumentalism. The wave-function is not a *sui generis* entity that exists in addition to the primitive ontology; but it exists, namely as being located in the particle configuration of the universe and its evolution.

In contrast to the dynamical parameters wave-function, mass charge, spin, etc., it makes no sense to say that a space-time in which the particle configuration is embedded or fields on that space-time are located in the particle motion. On a minimalist primitive ontology, there are only spatial relations that individuate sparse point particles; but there no space-time points. Hence, there are no field values occurring at space-time points either. Admitting these things to the ontology leads only to the commitment to surplus structure and the appearance of pseudo-problems as discussed in Sects. 1.3 and 1.5, in particular in view of the fact that fields in classical physics are not independent degrees of freedom given the distribution of particle masses and charges, as shown recently by Deckert and Hartenstein (2016) and Hartenstein and Hubert (2019). Space-time and fields in it are mere means of representation employed in a physical theory. In other words, dynamical parameters refer to the physical objects, namely to salient features of their motion, whereas the notions of space-time points and field values at space-time points do not refer to anything in the natural world.

The contrast between the dynamical parameters on the one hand and space-time as well as field values at space-time points on the other hand

---

[7] See for this view of the wave-function Miller (2014), Esfeld (2014b), Callender (2015) and Bhogal and Perry (2017). See Dowker and Herbauts (2005) for a concrete model in the framework of the GRW flash ontology.

shows up also in the fact that one can in principle dispense with the latter as means of representation. One can construct relational physical theories for both the domains of classical mechanics and relativistic gravitation as well as quantum mechanics that do not employ any absolute quantities (namely Barbour's shape dynamics, see Sects. 1.3, 1.6 and 1.7). Furthermore, one can construct a classical theory of electrodynamic interaction that does not employ fields on physical space-time in its formalism (i.e. the Wheeler Feynman theory, see Sect. 1.5). However, one cannot build a physical theory without using dynamical parameters. The reason is, as mentioned in Sect. 1.3, that no given configuration of matter as defined by the relative distances individuating basic objects contains information about the change of these distances. In order to formulate laws for that change, it is therefore mandatory to attribute some dynamical parameters to such a configuration that are defined in terms of their functional role for the change in the configuration and that are thereby located in that change. In brief, physical laws can dispense with absolute space-time and classical fields, but not with dynamical parameters that are defined in terms of their causal role for the evolution of the configuration of matter.

Over and above coming to an end when the salient patterns or regularities in the motion of matter are reached, scientific explanations also face a limit when it comes to the initial conditions of the universe. As mentioned in Sect. 1.4, in order to explain the observed increase in entropy, one has to admit what is known as the past hypothesis—namely that the initial state of particle motion of the universe (that is, the initial microstate) was a state that realizes an extremely low entropy macrostate of the universe. Given the low entropy initial condition and keeping the regularities of particle motion fixed as they are expressed in the laws of physics, one can then capture the entire evolution of the universe. However, one may wonder why there is such a low entropy initial condition. There are two types of reply possible to this concern within the scientific image.

The one type of reply is to argue in favour of lifting the status of the past hypothesis to the one of a law, because of its central position in an account of the universe that is simple and informative: the past hypothesis enormously simplifies our account of the universe and it contains a huge amount of information about what the evolution of the universe is like on a macroscopic scale.[8] As there is no explanation of why there are the patterns or regularities in the particle motion that there are *de facto*, so there is no explanation of why the initial state of the universe is a low entropy macrostate. Taking the past hypothesis to be nomological, because it simplifies the account of the natural world, makes intelligible why it does not call for an explanation in the same way as the fundamental dynamical laws do not demand an explanation.

However, taking the past hypothesis to have a lawlike status is not necessary in order to account for the direction of time, given the fact that the laws do not single out a direction of time. It is in dispute whether the past hypothesis is rich enough to account for the direction of time; instead, one can take the direction of time to be a primitive feature of time, if one endorses an absolute or substantival time, or—on relationalism—trace it back to change and take change to be directed as such (see Sect. 1.4 above). Also on these latter views it is intelligible why the laws do not single out a direction of them: doing so, they would lose simplicity without gaining in information about the world. Indeed, laws are most simple, when as they are fed with initial conditions as input, the yield both the future and the past evolution of the systems to which they apply as output.

The other type of reply to the quest for an account of the low entropy initial state of the universe is to seek to explain this special initial condition of the universe as we know it in terms of removing its peculiarity by enlarging the universe, that is, by maintaining that there is more to the universe than the evolution starting with this condition. Hence, the present state of the universe that we experience still has to be traced back to a particular initial state in the past; but the idea is that this state is not a special initial condition of the universe that has to be conceptualized in terms of a specific postulate as in the guise of the past hypothesis. Thus, in particular, Barbour et al. (2015) conceive the special initial condition to which we can trace the observable universe back as a turning point (so called "Janus point") in a larger evolution of the universe. The idea is that the overall evolution of the universe is such that it contains a turning

---

[8] See Callender (2004); and see Chen (2019) as to spelling this stance out in quantum physics.

point from which there is an evolution on both sides. One of these sides is the one that we know and that is commonly conceived as an evolution towards higher entropy states starting from a low entropy condition (conceptualized as the past hypothesis). However, Barbour et al. (2015) also maintain that this initial state is not a low entropy condition from which there is an evolution towards higher entropy states that will end in a state of thermal equilibrium of the universe. On the basis of relational mechanics, they argue that the evolution of the universe as a whole goes towards an increase of order in the sense of complexity instead of a decrease in the guise of higher entropy states; consequently, the turning point comes out as the condition with a maximum of disorder.[9] In any case, this is an explanation of the unification type. It shows how something that appears to be special is in fact integrated into a broader scheme. The apple falling from the tree is not peculiar, since it is an instance of the broader scheme of attractive motion like the Earth turning around the Sun, etc. By the same token, the special initial condition of the universe is not peculiar, since it is a particular element (the so called "Janus point") of a broader evolution.

However, one may wonder about the validity of that explanation in the present case. On the one hand, Barbour et al. (2015) give arguments how to integrate the evolution with a turning point into their programme of a relational ontology and dynamics (shape dynamics) that encompasses all of physics (see above Sects. 1.3 and 1.6). On the other hand, there is no possibility to test the hypothesis of a turning point, since one cannot go beyond the initial condition of the universe as we know it; there is no evidence of an evolution beyond that condition. More importantly and more precisely, for this reason, it is questionable whether giving such an explanation increases the coherence of the scientific image of the world: the coherence of that image is coherence with respect to all the available evidence.

In brief, as one cannot explain the laws that bring to the point the salient patterns in the motion of matter during the evolution of the

---

[9] For an account in a similar vein in terms of there being more than the observed universe, see Carroll (2010, in particular chs. 14–15). See Lazarovici and Reichert (2019) for a philosophical assessment of these accounts.

universe, so one cannot explain the border conditions for that evolution. For this reason, it is not objectionable to grant a lawlike status to the one border condition for the evolution of the universe that we know by admitting the past hypothesis (or whatever describes this condition) as nomological in virtue of its contribution to the scientific account of the world.

2.3   What Are Laws of Nature?

A primitive ontology of only matter in motion that takes everything else to come in through the way in which it enables a representation of the motion of the matter that is both simple and informative is associated with the view of laws of nature that goes back to the Scottish enlightenment philosopher David Hume. It is therefore known as Humean metaphysics or Humeanism in today's literature. Its central proponent is David Lewis. Lewis (1986b, introduction) is prepared to admit dynamical parameters such as mass and charge as intrinsic properties and also space-time and fields to the primitive ontology, namely to what he calls the "Humean mosaic", that is, the distribution of matter that makes up the universe. However, this leads to trouble when it comes to the wave-function in quantum physics; for the wave-function cannot be conceived as a parameter that is situated at space-time points or individual point particles.

The trouble for Lewis's Humean metaphysics that becomes evident when considering the quantum wave-function then triggered in today's metaphysics the elaboration of a stance that bans all these things from the "Humean mosaic"—that is, excludes the wave-function and, consequently, also mass, charge and classical fields as well as an absolute space-time from the primitive ontology, as worked out, in particular, by Elizabeth Miller (2014), Michael Esfeld (2014b) and Craig Callender (2015). The failure of Lewis's Humean metaphysics in quantum physics shows that something in this conception was misguided from the start: it was wrong-headed to admit to the primitive ontology (the "Humean mosaic") things that enter physics through their functional role for the motion of matter and its representation, as already sketched out by Ned

Hall (2009, § 5.2) as well as by Barry Loewer (2007). The primitive ontology consists only in those things for which these functional roles are exercised and that, consequently, can no longer be introduced in terms of functional roles for anything. These are point particles that are individuated by their relative distances and the change of these distances, that is, the particle motion. The resulting stance can be dubbed *Super-Humeanism*, as introduced and elaborated on in Esfeld and Deckert (2017, ch. 2.3). The qualification "Super-" expresses the attitude to ban everything from the primitive ontology that can be introduced in terms of a functional role and thereby be located in what really has to be endorsed as primitive. Super-Humeanism thus is the combination of a primitive ontology that is minimally sufficient to account for our scientific as well as common sense knowledge with Humeanism about laws of nature.[10]

On Humeanism, the particle motion comes first. This motion exhibits certain stable patterns or regularities as a matter of fact. The laws of nature as they figure in our scientific theories are our attempt to represent these regularities in such a way that we achieve an account of what happens in nature that is both as simple and as informative as possible. This account is known as the best system: it is the system that achieves the best balance between being simple and being informative. A system of only logical laws would be most simple, but not at all informative with respect to the actual universe. Dressing a long list of all the change that actually happens in the universe would be most informative, but not simple and parsimonious at all. Laws of nature seek to systematize this information.[11] They strive to be as simple as possible without losing the information about what happens in the actual universe. They are theorems of the system that strikes the best balance between these two factors.

More precisely, the idea is this one: Suppose that a description of all the relative particle positions and their change throughout the entire history of the universe is given. The laws of nature come out of this description as the theorems of a logical system that achieves an optimal balance between simplicity and information in the representation of these particle

---

[10] For critical discussions of Super-Humeanism, see Wilson (2018), Marmodoro (2018), Darby (2018), Lazarovici (2018), Matarese (2019) and Simpson (2019).

[11] See Hoyningen-Huene (2013) for an elaboration on systematicity as the central trait of science.

positions—modulo further parameters that are introduced by their functional role for the particle motion and that have to be specified as further initial conditions that enter into the laws. There can hence be laws of nature only if there are certain recurrent, stable patterns or regularities in the particle motion. But not any such regularity is promoted to the status of a law of nature: only those are that lead to an optimal balance between simplicity and informational content in the representation of the overall particle motion.

There is a well-known problem for Humeanism: the standards for what is simple, what is informative and what is the optimal balance between these two criteria may not be unique. Lewis's (1994) strategy to counter this objection is, in brief, to lay stress on natural properties that carve nature at its joints (such as, for instance, mass and charge). However, all the cognitive access that we have to these allegedly natural properties is the functional role that they perform for the particle motion, as also stressed by Jackson (1998, p. 23). Hence, if these properties had an intrinsic essence, it would be a pure quality that is beyond our cognitive access, as Lewis (2009) himself concedes. Consequently, the strategy that relies on such natural properties is in trouble already for purely metaphysical reasons, before it comes to the objection from the wave-function in quantum mechanics (and remember that in quantum mechanics, all the candidates for natural properties in physics—such as mass, charge, spin—are situated on the level of the wave-function).[12]

In any case, the strategy that relies on natural properties cannot be applied to Super-Humeanism, since there are no properties in the minimalist primitive ontology, but only relations, which individuate simple objects, and their change.[13] More precisely, there is exactly one type of natural relation that is the world-making relation, namely the distance relation. That is why it is not arbitrary on Super-Humeanism what the Humean mosaic is.[14] Nonetheless, in contrast to the predicates that are taken to pick out natural properties in Lewis's Humeanism, the predicates

---

[12] See Brown et al. (1995, 1996), Pylkkänen et al. (2015) and Esfeld et al. (2017) as mentioned in Sect. 2.1.

[13] See the objection that Matarese (2019) builds on this fact.

[14] Such an arbitrariness threatens, by contrast, the "package deal account" proposed by Loewer (2007) that is also directed against Lewis's natural properties.

that define the distance relation are not sufficient to formulate laws about the change in the relative distances among the material point objects. Further predicates have to be introduced in terms of their functional role for that change. But that is exactly the manner in which Lewis's Humeanism introduces the predicates that are supposed to refer to natural properties. However, it is immaterial for the significance that these functionally defined predicates have for the formulation of laws of nature whether one takes them to introduce dynamical parameters that are located in the particle motion through the described procedure, or whether one regards them as referring to natural intrinsic properties (to which we have no cognitive access anyway).

That is to say: in both Lewisian Humeanism and in Super-Humeanism, the procedure how to obtain laws of nature is the same. This procedure relies only on there being stable patterns in the particle motion. The patterns in the behaviour of the physical objects are what carves nature at its joints. These patterns enable the introduction of predicates in terms of a functional role for the particle motion. Laws of nature are then formulated in terms of these predicates. In both Lewisian and Super-Humeanism, it is an open issue whether this procedure leads to a unique best system of combining simplicity and informational content.

One may object to this procedure that physical laws do not describe only what happens in fact, but that they make also counterfactual propositions true. A physical theory not only informs us about the evolution of a given object provided that one fixes initial conditions, but it also tells us what *can* and what *cannot* happen with the objects in its domain. The theory thereby tells us also what we can do and what we cannot do in our actions.

However, making this fact intelligible only requires keeping the actual, salient regularities in the motion of matter fixed. Given these regularities insofar as they lead to the formulation of laws of nature in scientific theories, propositions about what can and what cannot happen with the objects that are covered by these laws have determinate truth-values. Thus, holding these regularities as represented in a scientific physical theory fixed, the state space of the theory, or the models that the theory admits, define what is possible and what is not possible for the objects in the domain of the theory. One may even speak of nomological necessity, that is, necessity with respect to the laws as fixed, provided that one bears in mind that, in Humeanism, the laws themselves derive from the patterns in the actual motion of matter. In other words, given these overall patterns or regularities, the truth-values for counterfactual propositions about what would happen with subsystems in the universe if they were in various initial conditions are fixed. This is all we are interested in when we handle particular systems.

The most important reservation against Humeanism about laws of nature is rooted in an intuition that is widespread also among scientists: science is not only concerned with representing the motions that occur in the universe in a manner that is both as simple and as informative as possible. In doing so, science discovers an underlying order of the universe. Hence, the intuition is that the salient observed regularities, which are expressed in the laws of nature that we formulate in our theories, are a manifestation of an underlying order of the cosmos. That order is modal, or even metaphysically necessary (i.e. could not have been otherwise). It constrains the motions that actually happen in the universe.

The intuition that drives Humean metaphysics, as stated succinctly in Lewis (1986b, introduction), by contrast, is this one: there is nothing modal in the world as such. In particular, there are no necessary connections. However, the project of the present book is not focused on that motivation. Its aim is not to advocate a particular position in the metaphysics of modality. Humeanism comes in here only as a metaphysical stance that enables us to bring out the following three aspects, which are central to the argument of this book:

1. In the first place, there is the distinction between primitive ontology and dynamical structure. Everything that can be introduced into a theory in terms of its functional role for the evolution of something is thereby located in that something. Hence, it does not belong to the primitive ontology. The latter only consists in those entities with respect to which the functional roles are defined and that, consequently, cannot be introduced in terms of a functional role for anything themselves. That is why they have to be endorsed as primitive. In the scientific image, these are simple objects (point particles) that are individuated by the distances between them and that move. The

position dubbed "Super-Humeanism" brings this distinction out. It is the key in order to understand why laws of nature are not opposed to human freedom, as will become clear in the next section.

2. Science seeks to identify patterns in the observed sequences of events such that laws can be formulated that simplify the representation of these sequences while still being as informative as possible about them and enabling predictions that can be tested. We can leave open here whether this is all there is to laws of nature, as according to Humeanism. The point is that this is all that explanations in science can achieve: explanations come to an end once the basic regularities in the motion of matter are discovered. Science cannot explain why there is matter and why there is motion at all. It cannot explain either why there is a universal pattern of attractive motion throughout the evolution of the configuration of matter of the universe. Science can only bring this pattern out in the guise of a universal law (i.e. the law of gravitation) and then explain through unification why particular phenomena of attractive motion are not astonishing. On this basis, science can then explain everything else in terms of causal roles for the universal patterns of motion and thereby locate it in these patterns.

3. The ontology of science consists in the answer to the following question: What are minimal ontological commitments that are sufficient to accommodate what science as well as our common sense knowledge tell us about the natural world? It does not lead anywhere to enrich the ontology of science with primitive modal entities that build up necessary connections, such as dispositions or powers that literally bring about or produce certain motions of matter, which are their manifestations. The same applies to the position of Maudlin (2007) that regards laws of nature as primitive and takes them to literally produce the motion of matter. One does not achieve an explanatory gain in that manner due to the mentioned circularity problem: the dispositions, powers or laws are defined in terms of the effects that they are supposed to bring about. One only runs into the impasse of pseudo-problems that have no solution, as discussed in the previous section: how does a disposition or power that is intrinsic to an object reach out of that object across space and make other objects move in a certain matter? How does a law, being an abstract entity if it exists, produce certain motions of matter?

One can employ Humean metaphysics in order to elaborate on these three aspects. The stress then lays on the transition to Super-Humeanism with its explicit distinction between a minimalist primitive ontology and the dynamical structure of physical theories. But we can leave open here whether what is expressed in these three aspects is all there is to the laws of nature or whether laws in nature pose some sort of general constraint or structure on how the particles can move without bringing about their actual motion.[15] Such an additional commitment may satisfy the above mentioned widespread intuition. However, one has to be clear about the fact that one does thereby neither achieve a gain in scientific explanations, nor an insight that could claim to be a deeper explanation.

2.4 Why Determinism in Science Is Not Opposed to Free Will

Suppose that classical mechanics is the correct physical theory of the universe. Then, given an initial state of the particle configuration of the universe at an arbitrary time and the laws of classical mechanics, the entire evolution of the universe is fixed by the laws and the initial state. That is, the entire *future* evolution from that state on as well as the entire *past* evolution leading to that state then is fixed. That is why the state at an arbitrary time can be used as initial state that is inserted into the laws. However, nobody within the universe can know its initial state at any time with enough precision to turn the determinism implemented in the laws into predictions (or retrodictions). But *determinism does not imply anything about the possibility of predictions*. It is a matter of the dynamical structure of a physical theory only: if—and only if—the laws plus the values of the parameters that enter into the initial conditions fix the entire evolution of the universe, then the universe is deterministic according to the theory under consideration.

If one considers physical determinism to be troublesome when it comes to human free will, a little reflection shows that the determinism

---

[15] See the reading of the historical Hume that Strawson (1989) proposes. Cf. also Esfeld and Deckert (2017, p. 56).

implemented in the dynamical structure of classical mechanics is not the source of the trouble. If there is trouble, the reason is the very fact of there being universal physical laws that, together with initial conditions, yield solutions of the equations in terms of which they are formulated. Suppose that a version of quantum mechanics that includes what is known as the collapse of the wave-function in its dynamical laws is the correct physical theory of the universe (such as the GRW theory, see Sect. 1.7). Suppose furthermore that the collapse of the wave-function is an irreducibly stochastic process. Nevertheless, the dynamical law then fixes objective probabilities for wave-function collapse to occur such that, given an initial quantum state of the universe at an arbitrary time that includes an initial wave-function of the universe, several possible future evolutions of the universe are fixed with objective probabilities attached to them.

If the decisions of human beings concerning the motions of their bodies influence neither the initial state of the universe nor the deterministic laws of classical mechanics, they do not influence the objective probabilities implemented in a fundamental stochastic law and an initial wave-function either, as argued by Loewer (1996) among others. Thus, in case the movements that persons actually make would systematically not fit the objective probabilities for these movements on the level of the particles that compose the human bodies, then the stochastic law would be refuted, as a deterministic law would be refuted in case persons do not make the movements that are fixed by the law and the initial conditions. Again, this is independent of the fact that nobody can make such predictions for lack of exact knowledge of the initial conditions that enter into these—deterministic or probabilistic—laws. Hence, supposing that there is a conflict between deterministic physical laws and free will, one could not draw any profit for free will if the laws were indeterministic. If there is such a conflict, it stems from the very fact of there being universal physical laws, be they deterministic or not.

The issue of such a conflict arises in the first place for universal laws. One may recognize laws also in the special sciences, in particular in the present context laws in genetics, evolutionary biology, or neuroscience. These laws may also be deterministic, so that one may consider a genetical, evolutionary biological, or neuroscientific determinism. But these are not universal laws. They hold only under normal conditions that cannot be exactly specified. For that reason alone they cannot pose a threat to free will. It is always possible that these laws fail because conditions are not normal. They investigate regularities that connect genes, biological evolution, neurophysiological configurations with the behaviour of organisms. However, since these are not strict regularities, there never is the situation that an exact specification of initial conditions plus these laws fix the behaviour of an organism.

The above mentioned, common formulation that, in the case of determinism, the laws plus the initial conditions fix the entire evolution of the objects to which they apply may be misleading. It may suggest that the laws somehow bring about the evolution of these objects. However, if this were so, the laws would have to bring about not only the *future* evolution of the objects from an arbitrary initial state on, but also the *past* evolution of the objects from an arbitrary initial state back. Deterministic laws do not single out a direction of time. But no one holds that the law brings about the *past* evolution by backwards causation. Hence, determinism as such contains no reason either to suppose that the law brings about the *future* evolution.

A better formulation of determinism that avoids any ontological connotations of the verb "fix" therefore is this one: the propositions stating the laws of nature and the propositions describing the state of the world at an arbitrary time (initial conditions) entail the propositions describing the state of the world at any other time. Thus formulated, it is clear that determinism in science is—only—about entailment relations among propositions. Supposing that determinism is true, the question then is this one: what it is in the world that makes these propositions true? In other words, in virtue of what in the ontology do these entailment relations among propositions hold?

One possible answer to these questions consists in maintaining that there are dispositions or powers in the world that literally produce or bring about and thus predetermine the evolution of the configuration of matter in the forward time direction through their manifestations, as argued notably by Alexander Bird (2007) and Stephen Mumford and Rani Lill Anjum (2011)—or that the laws themselves do this, as argued notably by Tim Maudlin (2007). In this case, the suspicion that laws of nature enter into conflict with human free will has a point: there then is something outside the influence of human decisions and actions that literally produces or brings about the motions of matter, including the motions of human bodies. Again, the conflict with free will is in this case independent of whether the operation of the dispositions, powers or laws is deterministic or not. If elements that lie outside of the influence of our decisions and actions fix objective probabilities that apply also to the motions of our bodies, then the suspicion is well-founded that there is a conflict with our free will.[16]

However, this is a specific view of laws of nature. It is not implied by the fact that universal laws figure in scientific theories. There are cogent objections against this view that are independent of considerations about human free will, notably the ones discussed in Sect. 2.2 above. Nonetheless, the issue of a conflict between laws of nature and free will is not settled by simply rejecting the view of laws, dispositions or powers as bringing about and thereby predetermining the evolution of the universe.

The most well-known statement of such a conflict is the consequence argument by Peter van Inwagen:

> If determinism is true, then our acts are the consequences of the laws of nature and events in the remote past. But it is not up to us what went on before we were born, and neither is it up to us what the laws of nature are. Therefore the consequences of these things (including our present acts) are not up to us.[17]

This argument can be construed as follows:

1. If determinism is true, then our acts are the consequences of the laws of nature and events in the remote past.
2. It is not up to us what went on before we were born.
3. It is not up to us what the laws of nature are.

---

[16] But see also the position that von Wachter (2015) advocates: according to him, it is wrong-headed to associate laws with regularities; by contrast, they indicate tendencies that can be trumped by the intervention of external factors such as, for instance, free will.

[17] Van Inwagen (1983, p. 16); for an earlier and more detailed version of the argument, see van Inwagen (1975).

4. From (1) to (3): The consequences of these things (including our present acts) are not up to us.
5. If our present acts are not up to us, we do not have free will.
6. Conclusion: Determinism implies that we do not have free will.

Again, what is at issue here is not determinism, but universal laws of nature, be they deterministic or not. If they are probabilistic, then the objective probabilities for our acts are the consequences of the laws of nature and events in the remote past. It then follows that if what fixes these probabilities does not include anything that can be controlled by us, we do not have free will. Furthermore, what is at issue here is not a view of laws that somehow produce or bring about the evolution of the configuration of matter of the universe. The term "consequences" in the consequence argument can simply be read as logical consequences in the above mentioned sense: given the propositions stating the laws of nature and the propositions describing the state of the world at an arbitrary time (i.e. the propositions describing the initial conditions), the conjunction of these propositions entails the propositions describing the state of the world at any other time. The entailed propositions include those ones that describe the state of motion of the particle configurations that are human bodies. Again, the fact that no one can deduce these propositions—and thus make deterministic predictions about the motions of human bodies—is irrelevant to the argument. Also the propositions about past states of motion are thus entailed. However, this does not touch upon the truth of premise (1). Its truth does not depend on any particular stance in the metaphysics of laws of nature.

One may consider endorsing premises (1) to (3) and dismissing premise (4), that is, deny that the laws of nature and the initial conditions of the universe on which we have no influence have the consequence that our present acts are not up to us. In this case, one defends the most widespread form of a compatibilism of determinism and free will; the most prominent such conception goes back to Harry Frankfurt (1971). The idea behind premises (4) and (5) is that if a person has free will, in those situations in which she exercises her free will, her acts are up to her. This means that she could have done otherwise. In other words, if, in a given situation, a person could not have done otherwise, her acts have not been up to her and she did not have free will in that situation.

Consequently, a compatibilism that rejects (4) cannot simply maintain that a person has free will, although she could not have done otherwise. The credibility of such a compatibilism depends on not simply rejecting the clause "could have done otherwise", but qualifying it without abandoning the determinism of laws of nature in terms of "could have done otherwise if the circumstances that led to her action had been otherwise". The problem with any such qualifications, however, is that, at the end of the day, in the framework of determinism they come down to saying that the person could have done otherwise only if the laws of nature or the initial conditions of the universe had been different. But this, then, just is what drives the reservations against this attempt to make free will compatible with determinism: however sophisticated an elaboration of qualifications of the clause "could have done otherwise" one may develop, it remains true on any version of such a compatibilism that our acts are the consequences of the laws of nature and the initial conditions of the universe, on which we have no influence according to (2) and (3).

It is certainly correct to point out that one cannot reject determination outright. Free will has to be distinguished from pure chance events in nature—that is, the idea of events that occur such that there are not even objective probabilities for their occurrence, because they do not come under any salient regularity. Thus, stressing that our actions are up to us or that, in exercising free will, persons could have done otherwise is not sufficient as a theory of free will. Any such theory has to spell out what distinguishes an action being up to us or a person's ability to do otherwise from pure chance events. I will do so in Sects. 3.4 and 3.5. That notwithstanding, if the qualification of the clause "could have done otherwise" cannot remove the implication that our acts are the consequences of the laws of nature and the initial conditions of the universe on which we have no influence according to (2) and (3), the intuition is well-founded that the baby of free will has been thrown out with the bathwater of pure chance events.

Hence, if the backbone of science, namely universal laws of nature, is not to contradict human free will, we have to attack premise (2) and / or premise (3) of the consequence argument. That is, we have to put forward a conception of laws of nature from which it follows that the laws and / or the initial conditions of the universe are in a certain sense "up to us". Furthermore, such a conception of laws has to be backed by cogent arguments that are independent of the issue of human free will. Moreover, it has to be situated within scientific realism: science discovers laws and initial conditions. They are not social constructions or in any way dependent on observers.

On Humeanism about laws, premise (3) is false. The reason is that first comes the motion of the matter in the universe, then come the laws as the axioms or theorems of the system that strikes the best balance between being simple and being informative in representing the motion that actually occurs. Our bodily motions are part of the motion of the matter in the universe, albeit only a very tiny part. Nonetheless, they are thereby part of the basis that determines what the laws of nature are. In that vein, for instance, Jenann Ismael writes:

> When we adopt a globalist perspective, our activities become part of the pattern of events that make up history. Since our activities partly determine the pattern, and the pattern determines the laws, our activities partly determine the laws. (Ismael 2016, p. 111; see also pp. 225–226)

In that sense, the laws are "up to us". Thus, if persons had chosen to do otherwise, the laws of nature would have been slightly different.[18]

This comptabilism of free will and (deterministic) laws of nature follows from Humeanism about laws only—that is, it follows from the claim that the laws are fixed by the mosaic alone, with our bodily motions being part and parcel of the mosaic. Consequently, this compatibilism is not tied to a particular theory of counterfactuals. It entails counterfactuals of the type that if persons had chosen to do otherwise, the laws of nature would have been slightly different; but it is not committed to a specific theory of counterfactuals. Hence, it is different from Lewis's (1981) own reply to van Inwagen's consequence argument, which, in turn, does not invoke Humeanism about laws, but rests on Lewis's theory

---

[18] See also Beebee and Mele (2002) for a detailed argument; and see already Swartz (2003, ch. 11, in particular p. 127).

of counterfactuals such that, in brief, a person having done otherwise would constitute a law-breaking event in the guise of a local miracle.

There are strong arguments in favour of Humeanism about laws of nature that are independent of the issue of free will. The most important argument is that this stance gives us all that motivates endorsing laws of nature in a scientific realist spirit—such as the significance of laws for explanations, for supporting counterfactuals, etc.—without going beyond a primitive ontology of matter in motion, that is, without calling for a commitment to primitive modal entities. But the drawback in the context of free will is that this conception seems to give us too much: there is a well-founded sense in which laws of nature define what we can do and what we cannot do. More precisely, we need laws of nature to delimit the range within which we can freely choose our actions. For instance, a person can choose to walk slowly or quickly, but she cannot choose to travel faster than light. By the same token, she can choose to walk left or right, but she cannot choose to jump as high as her house.

However, Humeanism has the means to take this difference into account: there are stable patterns or regularities of motion in the universe. Only if there are such stable patterns or regularities are there laws at all. Keeping these patterns or regularities as they are expressed in the laws of nature that we can formulate fixed, it is physically or nomologically impossible for a person to travel faster than light, or to jump as high as her house. This is all that is needed for the laws to define the range within which we can act freely. That notwithstanding, it is metaphysically possible that, tomorrow, a person travels faster than light or jumps as high as her house. There is nothing in the present or past configuration of matter that makes it impossible for such things to happen. It is only that if we presume that the salient regularities that were discovered in the past motion of matter will prevail in the future that we get to such restrictions. But there is nothing in the universe that makes it necessary that these regularities will prevail in the future.[19] Deliberating about her actions on the basis of the observed regularities of motion as expressed in the laws of nature that figure in our scientific theories is in any case all that a person can do, independently of whether or not there is something

---

[19] See again Beebee and Mele (2002, pp. 209–217).

in nature that constrains the future evolution of the configuration of matter.

Even if one concedes—*pace* the recent argument of Andreas Hüttemann and Christian Loew (2019)—that Humeanism can in this manner draw a distinction between metaphysical possibility and nomological or physical impossibility, the following impression remains well taken: regarding the laws of nature as being dependent on our actions comes into focus as a move in order to achieve the compatibility of our free will with the laws of nature only as a last means, precisely because the laws of nature delimit the range of what we can do.

Here again the virtues of Super-Humeanism show up: it is not only the laws, but the entire dynamical structure of the correct physical theory of the universe that depends on the change that actually happens in the universe. All the dynamical parameters that are introduced in terms of their functional role for the change in the primitive ontology—that is, the particle motion—are there to simplify, that is, to achieve a representation of the particle motion that is as simple and as informative as possible. Thus, they are not intrinsic to the particles or their configuration at any time. Whether or not a given particle or particle configuration realizes the role that functionally defines a certain dynamical parameter is a holistic affair: it depends on how the particle or the particle configuration in question moves within the whole particle configuration of the universe, as pointed out in comment (1) in the discussion of functional definitions in Sect. 2.1.

That is to say: the initial condition that enters as the state of the universe at a given time into the equations that express the laws of nature contains elements that are not intrinsic to what there is at that time, but that depend in the last resort on the overall change in the universe. In other words, they depend on the entire temporal evolution of the universe. These elements are notably the initial values of parameters such as mass, charge, fields, the universal wave-function, constants of nature. In order for these parameters to play their role to simplify the account of the motion that occurs in the universe, what role these parameters play and, notably, what has to be put in as their initial values, depends on the change that actually happens in the universe.

To stress again, the correct value of these parameters that enters into the state of the universe *at any given time* depends not only on what motion happens in the universe earlier than that time, but also on what happens *later* than that time. The reason is that these parameters are not located in the particle motion at any given time. They are located only in the overall particle motion throughout the history of the universe. Thus, to put it in a nutshell, we are ignorant of the initial wave-function of the universe at the initial state of the universe not only because of a principled limit on our knowledge, but also because what is the *initial* wave-function of the universe is only fixed at the *end* of the universe.

Hence, human actions influence in the first place the initial values of the dynamical parameters that are part and parcel of the initial state of the universe and that enter into the physical theory in terms of their functional role for the evolution of the configuration of matter of the universe. Therefore, if human beings chose to do otherwise, in the first place, slightly different initial values for the dynamical parameters at the initial state of the universe would have to be figured out in order to achieve a system that maximizes both simplicity and informational content about the change that actually occurs in the universe.

For the sake of illustration, assume that quantum mechanics is the correct theory of the universe. Then, what would be slightly different if humans chose to do otherwise than they actually did would not be the Schrödinger equation and / or the Bohmian guiding equation or the GRW collapse law, but the initial quantum state of the universe. That is, the values that the universal wave-function takes as initial condition in the configuration space of the universe would be slightly different in that case. Recall that the quantum state is not the state as given by the primitive ontology (such as particle positions) at a time, but the state as given by an initial wave-function. However, what the wave-function is at a time depends on the entire evolution of the configuration of matter (the primitive ontology). The wave-function is a dynamical parameter that is defined by its functional role for the evolution of matter and thereby located in that evolution as a whole (see Sect. 2.2).

In that way, van Inwagen's consequence argument turns out to be invalid without the Super-Humean being committed to saying that it is up to us what the laws of nature are. Instead, there is an ambiguity in the phrase "it is not up to us what went on before we were born": this statement may mean that we cannot change what happened in the universe before our birth. In that sense, the Super-Humean endorses this statement. But this statement can also include reference to an initial state of the universe before we were born. Insofar as that initial state enters into laws of nature, it includes values of parameters that are not intrinsic to it. These values depend on what happens later in the universe, including the particle motions that are expressions of human free will. Van Inwagen has to use this statement in that latter sense. Otherwise, we would not get to steps (4), (5) and (6) of the argument. The past particle motions have consequences for the future, including our present acts, only if they include values of additional dynamical parameters that enter as initial conditions into the laws.

Consequently, rejecting premise (2) does not imply that human decisions about how to move their bodies alter past particle positions and motions. This distinguishes the present proposal from proposals in terms of backwards causation, such as the one of Peter Forrest (1985). The particle positions and motions before we were born are what they are, independently of what we do. All observations, including all measurement records, are position observations and are recorded as spatial configurations (see Sect. 1.1). Human free will does not alter past observations or touch upon the validity of records of the past. However, there is a distinction between position as primitive parameter and the additional dynamical parameters that enter into the initial conditions for a law of motion. These latter are functionally defined in terms of the role that they play for the motion that actually occurs. Their initial values can therefore be dependent on future motions, including motions that are the result of free will, without any paradox arising. They are located in the motion as a whole.

Again, this proposal does not depend on a particular theory of counterfactuals. It only exploits the idea according to which the dynamical parameters that define a state that enters as initial condition into a law depend on the motions that actually happen in the universe, including the motions of human bodies. Of course, this proposal then implies counterfactuals to the effect that if persons had chosen to do otherwise, the initial values of some of these parameters at the initial state of the universe would have been slightly different; but it remains neutral on the theory of counterfactuals.

This proposal is also distinct from the way in which premise (2) is rejected in proposals that are situated within the metaphysics of a block universe and that are combined with Humeanism about laws, so that premise (3) comes out false as well.[20] In these proposals, premise (2) is dismissed only because (i) all events exist in a timeless manner and (ii) determinism implies that, given the complete description of the state of the universe at an arbitrary time—which may be a time at which persons live and act—plus the laws, also the propositions about all the *past* states of the universe are entailed. Thus, also on the block universe metaphysics, it can be true that if humans had done otherwise so that the state of the universe at a certain time would have been different, the initial conditions of the universe would have been different as well. But this is not sufficient to make the block universe metaphysics compatible with free will; it does not imply that humans actually had the possibility to do otherwise.

The problem with the block universe metaphysics is that there is variation in the block universe, but no change (see Sect. 1.6). However, without change, there is no possibility of free will. The issue of free will is not one of static dependency or determination relations among timeless events, but whether there is a free will that makes it that certain rather than other changes happen (such as me having coffee instead of tea for breakfast). However, if there is no change, the question does not arise whether some of the change that happens or can happen is "up to us". The—usually—tacit presupposition that without change, there would be no question of free will, is at the roots of employing a conception of determinism in the debate about free will according to which initial conditions plus the laws determine the *forward* evolution of the configuration of matter. Nonetheless, those that use such a formulation are aware of the fact that determinism in physics does not single out a direction of time. Hence, *pace* Hoefer (2002) and Ismael (2016, ch. 6), the block universe metaphysics combined with Humeanism about laws does not

---

[20] For such proposals, see Hoefer (2002) and Ismael (2016, ch. 6 and pp. 227–230). See Brennan (2007) for a criticism of Hoefer's proposal.

amount to a rejection of premise (2) that makes room for free will, since there is no change on that metaphysics.

By contrast, the present proposal admits change as a primitive. In consequence, in the framework of this ontology, premise (2) is rejected in a differentiated manner: the values of *only some* of the parameters that enter into the initial conditions of the universe according to a physical theory depend on the actual change that occurs in the universe, namely those that are defined in terms of their functional role for the evolution of the primitive parameters. But this does not hold for the values of the parameters that make up the primitive ontology. We thereby obtain a sense in which a part of the initial conditions of the universe that enter into laws of nature is "up to us" in a manner that is pertinent to free will: the values of the former parameters did not yet exist when the initial state of the universe occurred long before we were born, because they are not intrinsic to that state, but are located in the entire motion of the matter of the universe.

A similar consideration applies to the laws in the special sciences. Again, even leaving aside the fact that if these laws were deterministic, they would always apply only against the background of normal conditions, a genetical, evolutionary biological or neurobiological determinism would only hold on the basis of specifying appropriate initial conditions. However, parameters enter into these initial conditions that are defined by their biological or neurobiological function. Consequently, one can apply the procedure of location also to the dynamical parameters that figure in the laws of the special sciences and that are defined in terms of their functional role for the evolution of the systems to which they relate. These parameters then are located in the evolution of these systems. Their initial values thereby depend on the actual evolution of these systems. Hence, it follows that some of the values of parameters that enter into these initial conditions depend on the actual behaviour of humans in the just mentioned sense: if humans had chosen to do other things than they actually did, some of the initial values of these parameters applying to states that occurred before the human actions in question happened would have been slightly different.

Furthermore, on Humeanism about laws, special sciences' laws are again just a pattern in the phenomena, instead of something that governs, produces or brings about the phenomena. Thus, as a matter of fact, human decisions and bodily motions may manifest certain stable regularities with respect to genes, neuronal configurations, etc. Nevertheless, the decisions and bodily motions come first and then come the regularities that may connect genes, neuronal configurations, etc. with decisions.

Premise (3) of the consequence argument—the laws of nature are not up to us—comes out false in a scientific realist framework only if one adopts Humeanism about laws. For premise (2)—the initial conditions are not up to us—to be falsified without endorsing backwards causation, one has to make the step from Humeanism to Super-Humeanism. One has to reject notably the idea that the dynamical parameters refer to intrinsic properties of the systems to which they are attributed. However, premise (2) can turn out to be false also if one rejects Humeanism about laws. It is sufficient for premise (2) to be proven wrong that one endorses what Super-Humeanism says about the dynamical parameters. For that, one only has to subscribe to the definition of these parameters in terms of their functional role for the evolution of the elements of the primitive ontology. More precisely, one has to regard these definitions as being such that they locate these parameters in the primitive ontology, such as the particle motion that actually occurs. If these parameters are thus located, they are nothing over and above the particle motion. Consequently, their determinate values are not yet there at the initial particle positions and their change. They come out of the particle motion.

This stance can therefore also go with an anti-Humean view of laws of nature according to which the laws are ontological primitives that pose a general constraint on how the matter in the universe can move without actually producing the motion of matter. One may prefer such a view of the laws in order to anchor the scope that the laws of nature define for our actions in the ontology. What a person can and what she cannot do is any case fixed by the laws of nature, be they ontological primitives or be they reducible to the salient regularities of the motion of matter that hold throughout the universe.

The point of rejecting premise (2) rather than premise (3) of the consequence argument, whatever stance one takes in the metaphysics of laws, is this one: taking determinism for granted, whether a person, for instance, goes left instead of right, does not depend on the laws as such, but on the initial values of the dynamical parameters over and above those defining the primitive ontology that enter as initial condition into the laws; and these values, in turn, are not fixed before the person decides to go left or right, but depend on her decision among other things (unless one considers the dynamical parameters as intrinsic properties—as on Lewis's Humeanism—or as dispositions, powers or structures that sit in the objects and thereby runs into the drawbacks discussed in the previous section that are all independent of the issue of free will).

In sum, Super-Humeanism leads to an original rebuttal of van Inwagen's consequence argument and thereby to a new version of a compatibilism of deterministic laws of nature and free will, which is based on a differentiated rejection of the premise according to which the initial conditions of the universe are not up to us. By dismissing this premise, the Super-Humean proposal does not come into conflict with the well-founded widespread view according to which the laws delimit the range within which we can freely choose our actions. More precisely, what a person does in a given concrete situation (or what are the objective probabilities for what she does) is not determined by the laws as such, but by the values of the dynamical parameters that enter into the initial conditions. Super-Humeanism addresses precisely this point in taking these values to be dependent on the overall evolution of the configuration of matter of the universe, which includes the human bodily motions. This is an important advantage in comparison to the compatibilism within standard Humean metaphysics (Humeanism about laws), which rests only upon the premise of the laws being dependent also on the human bodily motions and does simply not address the issue of the initial values. Consequently, the Super-Humean reply to van Inwagen's argument also is available in case one has reservations about the Humean rejection of any sort of constraint that the laws put on the evolution of the configuration of matter of the universe.

Actions resulting from free will come nevertheless under regularities. There is a systematic connection between human intentions, changes in the brain and bodily motions that can be tested in psychological and neurobiological research. If there are regularities that connect free decisions with bodily movements, then we have to answer the following questions: Why do such regularities not show up in the form of particular dynamical parameters for free will that, in the last resort, would have to be included in the laws of physics? If free will is distinct from chance events, why does it not manifest itself in the laws?

Science strives for laws that are as general, as simple and as informative as possible. But it depends on the actual motion of matter to what extent such laws can be achieved. Consequently, it is by no means mandatory that the same simple laws—such as the law of gravitation—apply to all the particle configurations and all the change in the universe. Thus, it may be so that at some level of organisation in complex systems, the particle interactions manifest regularities that are incompatible with the basic physical ones. Carl Gillet (2016) even claims that there is scientific evidence for this and speaks of strong emergence and downward causation. If this were so, hence, particle interactions in brains may manifest regularities that deviate from the basic physical ones due to human free will. Nevertheless, in the ontology, there are still only point particles and their configurations that move, although the regularities of motion become different when configurations reach a certain degree of complexity.

However, *pace* Gillet (2016), there is no established evidence that the basic physical regularities as expressed in the laws of physics break down in some complex physical systems. The neuroscientific research that we know is applied physics—applied classical mechanics and electrodynamics, or, maybe, quantum mechanics in case quantum effects are proven to be operational in the brain. More importantly, also when it comes to free will, there is no need for any such breakdown to establish that free will is compatible with what science tells us about the world.

The search for such a break down is driven by a misunderstanding of laws of nature and scientific research. Science strives for systematicity.[21] For scientific laws to be general, simple and informative, they better not include dynamical parameters that refer to the behaviour of only some specific objects in the universe—unless one were to lose the information about the evolution of these objects without such additional parameters. The evidence on the basis of which one formulates general, simple and informative laws contains all the motions that occur in the universe, including the motions of human bodies that enjoy free will. However, the free will is irrelevant for that evidence: in science, all the evidence consists

---

[21] See again Hoyningen-Huene (2013).

only in the motions of bodies. What is relevant for the issue of free will only is this: as mentioned several times, laws merely pose general restrictions for motions. They set a frame for the motions that are dynamically possible. One obtains propositions about specific motions only if one feeds the laws with initial values of dynamical parameters. These parameters are defined in terms of their functional roles for these motions. They are thereby located in these motions. That is why determinism in science does not *pre*determine any motions that actually occur. Consequently, there is no conflict with free will.

The initial values of these dynamical parameters always vary slightly as soon as one varies motions in the universe. Hence, the initial values that these dynamical parameters take as a matter of fact are different in a universe with exclusively inorganic matter from what they are in a universe in which some complex physical systems are organisms and a universe in which some complex physical systems are humans with free will. Recall that there are purely physical sufficient conditions for particle configurations to be organisms including human beings. Indeed, nothing what has been set out in this chapter is specific for persons and their free will. What is at issue here only is how the evidence in the guise of all the motions that occur in the universe is related to the laws and the dynamical parameters for which initial values have to be set in order to apply the laws to specific motions. Analysing this relationship is sufficient to remove the concerns that one may have about laws of nature coming into conflict with free will.

This chapter and the preceding one have removed two issues that are widely considered to be focal points of a clash between science and the common sense view of us as persons: certain questions about time and free will. As regards time, science does not oblige us to endorse the ontology of a four-dimensional, spatio-temporal block universe: even the physics can be formulated without the geometry of such a block universe. As regards free will, the standard argument for an incompatibility between free will and deterministic laws in science (more generally speaking, universal laws) fails, because the laws require initial conditions to operate. Into these initial conditions enter parameters that can be taken to be located in the motion of matter, because they are functionally defined in terms of their role for the motion of matter; that motion includes human behaviour.

The issues of time and free will are obviously connected. What connects them is that change is required for free will to operate. Furthermore, what change occurs has to be open so that some of the change that comes about depends on the free will of persons. Against this background, the solution that this book proposes to both the issues of time and free will is this one: put change as an axiom into the primitive ontology, for it cannot be derived from anything else; restrict the primitive ontology to what is minimally sufficient to account for our scientific as well as our common sense knowledge, for this is the way to avoid an ontological commitment to surplus structure and the problems that such a commitment entails. As has turned out in this chapter, the issue of a clash between free will and determinism (more precisely, universal laws in physics) is one such pseudo-problem (*Scheinproblem* to use the term of Carnap (1928)).

To stress again, both the problems of time and free will arise only if one enriches the ontology of the physical world beyond what is minimally sufficient to account for the evidence in a scientific realist spirit, namely in the former case to an absolute space-time in the guise of the four-dimensional geometry of the block universe and in the latter case to something modal that is intrinsic to the initial state of the universe and that *pre*determines the evolution of the universe. To put it in a nutshell, if properly spelled out by using minimal sufficiency as a guideline, the ontology of science turns out to be not rich enough for a problem of time or a problem of free will to arise. Hence, we get as benefit from the parsimonious primitive ontology set out at the beginning of this book that the alleged clashes between science and the common sense view of ourselves as persons with respect to time and free will are removed.

This chapter and the preceding have not gone beyond dispelling these clashes. There is no conception that is specific for human free will forthcoming from the considerations in this chapter. All that has been said here about the motion of objects determining the values of dynamical parameters that enter into initial conditions (as well as the laws, on Humeanism about laws) applies to all the objects in the universe, electrons and humans alike. We will go into a specific conception of persons and their free will in the next chapter after having spelled out the real issues of the conflict between the scientific account of the world and the view of ourselves as persons.

3

Why the Mind Matters: The Manifest Image of the World

3.1 Sensory Qualities as Problem for the Scientific Image

The scientific image of the world, insofar as it has been developed by modern science, can be traced back to René Descartes, who formulated the conception of nature as *res extensa*.[1] This conception can then be spelled out as the primitive ontology of distance relations individuating featureless particles and the change in these relations (that is, the particle motion). Descartes did not consider the scientific image to be complete. The mind, conceived by Descartes as *res cogitans*, is situated outside this image. Moreover, everything that is not obviously construed in terms of the motion of physical objects is relegated to the mind.

Consider the case of colours. For Descartes and the whole mainstream of early modern philosophy, colours are a paradigmatic example of what is known as secondary qualities in that stream. They belong to the mind and thus stand in contrast to primary qualities such as the geometrical properties, which belong to the objects. Not only colours, but all the

---

[1] See notably Descartes, *Principles of philosophy*, part 2, § 5.

© The Author(s) 2020
M. Esfeld, *Science and Human Freedom*,
https://doi.org/10.1007/978-3-030-37771-7_3

sensory qualities—including sound, smell and taste—are considered to be secondary qualities and situated in the minds of perceiving subjects.

This classification faces a serious objection: predicates such as "red", "noisy", "stinking", "sweet", etc. do not apply to anything mental. There is nothing red, nothing noisy, nothing stinking and nothing sweet in the minds of persons. These predicates apply exclusively to extended, physical objects. It may be that these predicates do not denote intrinsic properties of particle configurations. It may be that particle configurations are coloured and emit a certain sound, smell or taste only in relationship to other particle configurations that count as perceiving subjects. Nonetheless, colours, sounds, smells, tastes then are certain relations among particle configurations. Hence, they belong to the physical world.[2]

However, it is not obvious how to locate these qualities in the scientific image of the world. Their case is different from the one of the parameters that do not belong to the primitive ontology and that make up the dynamical structure of a scientific theory. We have access to parameters such as mass, charge, energy, wave-function, etc. only through the scientific theory that introduces them. We do not observe these parameters, but only the relative positions and motions of discrete objects. Accordingly, the dynamical parameters of scientific theories are defined in terms of their functional role for the particle motion. They are thereby located in the particle configuration of the universe and its change. Consequently, the description of the particle motion entails the propositions that attribute these parameters to physical objects (see Sect. 2.1).

By contrast, our access to colours, sounds, smells, tastes is prior to science. It is at least at first glance not clear how this procedure of location through functional definition can apply to them. The sensory qualities certainly are important to distinguish macroscopic objects from one another. But this does not mean that the perceived differences among them can be accounted for in terms of a functional role for the motion of matter. Furthermore, at least in the case of colours, a problem of permutation arises, as pointed out by Michael Smith (1994, pp. 48–50): even if there were a functional difference between, say, red and green, it seems that one could permute the ways in which objects appear as red and green

---

[2] See e.g. Jackson (1998, ch. 4) for such an account.

to persons without touching upon these functional roles (that is, switch what appears as red to appearing as green and *vice versa*). In other words, all that matters for these functional roles is that objects have *some* colour or other by which they distinguish themselves from one another. However, it is of no importance which colour this is. If this is so, the description of the particle motion cannot entail which objects are red, which ones are green, etc. By contrast, one could obviously not permute the ways in which gravitational and electromagnetic motion appear to observers.

The problem how to locate the sensory qualities in the scientific image of the world is independent of which primitive ontology one endorses. Thus, it is independent of maintaining a primitive ontology that is committed only to point particles and their motion or enriching the ontology such that it contains masses, charges, forces, fields and the like as further primitives. If one subscribes to a minimalist ontology, one endorses only particles that move, but no fields as mediators of their interaction (see Sect. 1.5). If there are no fields, there is no light either. The phenomena of electromagnetic radiation, including stars emitting light, are accounted for in terms of direct particle interaction only. This is not absurd: recall that, as mentioned in Sect. 1.1, if a theory gets the spatio-temporal arrangement of the particles—more precisely, the fermions—right, it has got everything right that can ever be checked in scientific experiments; consequently, two theories that agree on the spatio-temporal arrangement of the particles cannot be distinguished by any empirical means, whatever else they may otherwise say and disagree on. Hence, there is nothing empirically incoherent in a theory that rejects light.

Rejecting fields does not make the problem how to locate colours in the scientific image of the world more difficult to solve than if one admits the electromagnetic field as a further primitive along with the particles. In the latter case, the standard strategy is to identify colours with wavelengths of electromagnetic radiation and how this radiation is reflected by the surfaces of certain particle configurations. In the former case, the one of a direct particle interaction theory, the corresponding strategy is to identify colours with certain interactions among particle configurations that lead to certain accelerations of particles. In neither case come colours out as intrinsic features of physical objects. In both cases, they are located in interactions.

In both cases, the problem is the same: how can electromagnetic radiation of certain wave-lengths reflected by surfaces *be* blue, red or green? How can certain interactions among particle configurations *be* blue, red or green? Including particle configurations that are perceiving subjects in these interactions does not make this problem easier to solve. Correlating perceived colours with certain wave-lengths of electromagnetic radiation or with certain particle interactions, and be it particle interactions that include the brains of perceivers, is not sufficient to solve this problem, however systematic such correlations may be.

Only a functional definition of colours in terms of their role for particle motions could do so, following the model of how dynamical parameters such as mass, charge, forces, energy are introduced in physical theories in terms of their causal role for particle motion and thereby located in that motion. Such a definition then makes intelligible on the basis of their motion how some particles can be electrons, protons, neutrons, etc. Accordingly, only a functional definition in terms of a causal role for particle motion would make intelligible how electromagnetic radiation of certain wave-lengths or certain particle interactions can *be* blue, red or green, etc. Let us therefore note the issue *how to account for sensory qualities as the first big problem for the scientific image as the complete image of the world*.

This issue is usually conceived as the problem how to integrate sensory experiences such as seeing, hearing, smelling, tasting, touching as well as states that are not directly related to one sense organ such as feeling pain, being in love, etc. into the scientific image. All these are characterized by a certain qualitative feeling—what it is like to be in love, in pain, to see the colour of ripe tomatoes, to listen to a musical performance, to smell hot coffee, to taste a good wine, etc. They are therefore known as *qualia*. To account for them is considered to be the hard problem of consciousness, as brought out notably by David Chalmers (1996, introduction). However, this is not the entire problem: as mentioned above, predicates such as "red", "noisy", "stinking", "sweet", etc. do not apply to anything mental, but to objects in the world. The problem hence concerns not only the states of consciousness known as *qualia*, but how to locate sensory qualities as a whole in the scientific image.

Sellars (1962, end of section VI) puts forward the speculation that new physics is called for to solve this problem. In later writings, he elaborates on this idea in terms of an ontology of absolute processes and a science based on such an ontology.[3] The processes are absolute because they are primitive: they are not sequences of events, which, in turn, consist in the change of properties of or relations between objects. There are no objects in this ontology. It is purely Heraclitean in the sense that there only is change, but nothing that changes. Interestingly enough, remarks in a similar vein can be found in the—esoteric—writings of the later Bohm.[4] Bohm there seeks to dissolve objects into pure processes.

Suppose that an ontology of pure processes could be spelled out in a clear and coherent manner and that a scientific theory of physics could be based on it that has the same explanatory power as the physics that we know. Thus, instead of making statements such as "Particles attract each other in a gravitational manner", "Particles attract and repulse each other in an electromagnetic way", "Particles move electronwise", etc., one would have to say "It gravitates", "It electromagnetizes", etc. One would have to develop a geometry at least as a representational means in order to specify where and when it gravitates or electromagnetizes.

However, even granting the possibility of such a physics for the sake of the argument, it remains unclear what could be the gain for solving the problem how to account for sensory qualities within the scientific image. Even if the scientific image were cast in terms of a physics of absolute processes, some processes would have to be singled out as basic in the sense of making up the primitive ontology. These would be processes of gravitating and electromagnetizing. Reddening, or pinking ice-cubishly (the example of Sellars 1962), would in any case not be candidates for basic processes.

Hence, the mentioned problem would still be with us: instead of locating colours, sounds, smells and tastes and their sensory, qualitative experience—conceived as features of objects—in the scientific image construed in terms of basic physical objects in motion, the task would be

---

[3] See notably Sellars (1981). See Seibt (1990) for pursuing a process ontology. Strawson (2017) combines a process ontology with panpsychism.
[4] See the last chapter of the—serious—book Bohm and Hiley (1993, ch. 15); and see Bohm (1980). See Pylkkänen et al. (2015) for an analysis that seeks to take these remarks seriously.

to locate these features—conceived as absolute processes—in the scientific image construed in terms of basic physical processes such as gravitating and electromagnetizing. The latter task is as intricate as the former one. To put it differently, one may alter the content of the scientific image; that alteration may even go as far as replacing one ontological category (such as the one of objects) with another category (such as the one of pure processes). But the methodology of metaphysics would remain the same, namely the one sketched out in Sect. 2.1 above. Thus, if one endorses the scientific image, the problem how to locate in the ontology of this image what does not figure explicitly in its basic notions would remain the same, whatever the proposed ontology is.

By calling for new physics in order to solve the problem how to locate the sensory qualities in the scientific image of the world, Sellars dissociates himself from the tradition in modern philosophy that is associated with Descartes and that distinguishes between primary and secondary qualities. Sellars thereby turns back to ancient philosophy. Before Descartes put perception and the features of physical objects to which we have direct access only through one particular sense organ in the mind, both these features and their perception were considered to be part of the natural, physical world. Only the mind in the sense of reasoning (*lógos*) was regarded as distinct from the physical world.

One can consider Descartes' position as being motivated by the problem of location: if one cannot locate the sensory qualities in the scientific image of the world as conceptualized in terms of *res extensa*, one can deny that they are properties of physical objects. That is to say, one can pursue the strategy to eliminate them from the scientific image of the world, if one assumes that it is not possible to reduce them to the basic notions in terms of which this image is formulated. However, one cannot deny that physical objects *appear* to persons as being red or blue, quiet or noisy, stinking or smelling pleasantly, sweet or sour, etc. Hence, the problem is shifted from accounting for colours, sounds, smells and tastes as features of the physical world to accounting for their appearances in the minds of persons. Obviously, pursuing this strategy makes sense only against the background of assuming that the mind is distinct from the physical domain. Only in this case can one relegate everything to the mind that cannot be introduced in terms of its functional role for the motion of matter. Obviously as well, the mind, *res cogitans*, then is as primitive as matter, *res extensa*.

In contrast to what Descartes took for granted, today, the progress of science does not stop at the mind. Neuroscience has the ambition to turn the mind itself into an object of scientific investigation. But then the following conclusion imposes itself: if the mind admits of a scientific investigation, then all the features that belong to the mind are part and parcel of the physical domain and are hence located in matter in motion. Consequently, there is no point in banning whatever features from the physical domain and relegating them to the mind. More precisely, one makes the problem even harder to solve, if the strategy to relegate features to the mind that do not figure explicitly in the scientific image of the world is excluded because everything is subject to scientific investigation. The challenge is that if one takes scientific investigation to have no limits even when it comes to the mind, one thereby loses also the possibility to account for the *appearance* of a world that is coloured, in which there are sounds, smells and tastes. Hence, one may deny that colours, colours, sounds, smells and tastes are properties of physical objects; but in doing so, one must be sure not to lose the means to account for the appearance of these features to persons. However, to locate this appearance in the ontology of the scientific image is at least as hard, if not harder than, directly locating colours, sounds, smells and tastes in matter in motion.

In case one cannot locate these features in the ontology of the scientific image through functional definitions, there still remains the option of simply stipulating that colours, sounds, smells and tastes as well as their sensory, qualitative experience are identical with particle motions. The argument then is that the identity theory still comes out as the best option in comparison to dualist accounts on the one side and eliminativist accounts on the other side.[5] However, in this case, one loses all the explanatory power that the method of location through functionalization provides. There then is no answer to the question why certain particle motions in the brain are conscious experiences. This is a brute fact. By contrast, it is reasonable to demand to do better and give an explanation of the features of the world that do not figure explicitly in one's primitive

---

[5] See e.g. Papineau (2002) for endorsing that option.

ontology, whatever primitive ontological commitments one subscribes to. Simply stipulating that all the rest is identical with elements of the primitive ontology that one endorses merely makes evident that one has no solution to the problem of how the features that do not figure explicitly in one's primitive ontology come into the picture.

This loss of explanatory power is made evident by the term "explanatory gap" coined by Joseph Levine (1983). There are correlations between brain states and conscious states that may turn out to be as strong or systematic as one likes, and yet, following Levine, it is not intelligible how brain states can *be* conscious states. According to Levine, this problem arises only in the case of brain states and mental states. It does not come up, for instance, in the case of H_2O molecules and water. Water can be functionally defined in terms of its thirst-quenching role. Science identifies H_2O molecules as being under standard environmental conditions the particle configurations that fulfil this role. Consequently, H_2O configurations *are* water.

However, one can object to Levine that the explanatory gap, if it comes up at all, is rather a matter of degree than the one of a clear-cut separation between cases in which it does not arise (such as H_2O molecules and water) and a case in which it arises (such as brain states and conscious states). The issue is to what extent one is convinced that the scientific image provides an account of what there is in the world and when (if at all) one abandons this conviction. Setting out functional definitions in terms of a causal role for particle motion is just a matter of definition, as pointed out in comment (2) in the discussion of functional definitions in Sect. 2.1. Such a definition can be formulated for whatever comes up for location in the scientific image of the world, because it does not figure explicitly in this image as given by the basic notions that describe matter in motion. By the same token, as soon as one goes beyond the dynamical structure of a physical theory with parameters such as mass, charge, energy, wave-functions, etc. that are indispensable in order to formulate laws of motion for the material objects as defined by the basic notions, one can raise doubts whether this procedure of location through functional definitions is convincing.[6]

---

[6] See in particular the argument of Brandom (2015, pp. 80–85, 231–235) for raising such doubts.

Indeed, one could in principle voice such reservations already in the case of water, that is, doubt whether the motion of H_2O molecules is all there is to water. However unreasonable such doubt may be, the point is this one: the scientific theory in chemistry that treats H_2O molecules does not qua scientific theory of H_2O molecules commit us to accepting that *water* is identical with configurations of H_2O molecules. Water could be more than just configurations of H_2O molecules sticking together under standard environmental conditions. It is in principle possible to maintain that as mental states are not brain states, although both are strongly correlated, so water is not H_2O molecules, although both are strong correlated. It is in both cases an issue of endorsing a functional definition of mental states or water in terms of a causal role for particle motion or not. It is difficult to see a cogent reason not to do so in the case of water, whereas this is much less difficult in the case of mental states.

Thus, the issue of an explanatory gap is a gradual one: there is little reason to admit an explanatory gap in the case of water, probably a bit more reason in the case of organisms and evidently a reason to do so in the case of conscious experience. However, that evidence may be misleading: by the same token, there is nothing that commits one to admit an explanatory gap at all. One can also maintain that everything, including sensory qualities and conscious experience, admits of a functional definition in terms of a causal role for particle motion. Future neuroscience may do for sensory qualities what chemistry did for water and molecular biology for organisms.

We can draw two conclusions from these considerations: (1) It is an open issue whether there is a principled problem when it comes to sensory qualities and conscious experience that cannot be solved within the scientific image. The arguments that there is such a problem rest on the intuition that these are purely qualitative features, which, consequently, do not admit of a functional definition; but this intuition may be misleading. (2) If there really is a problem, then it is not limited to the mental: colours, sounds, smells, tastes are not situated in the mind, but in the world, even if they may turn out to be relational features. At least this is brought out by Sellars's speculations about new physics in the guise of a process ontology, however unconvincing that ontology may be. Consequently, if one thinks that the scientific image of the world fails when it comes to sensory qualities, one has not obtained a convincing physicalism, or something near enough, *pace* the title of the book of Jaegwon Kim (2005).

3.2   Normativity as the Focal Point

Conscious experience is not the feature that distinguishes us humans. Some non-human animals also have conscious experience. What distinguishes humans is thought. This is so because thought—and thought only—has meaning. That is, it is directed towards something: it refers to something and says something about its referent or purported referent. Let us therefore consider now the relationship between human thought and the scientific view of the world.

One way to enter into this discussion is to start from artificial intelligence. Syntactical relations in the sense of regular sequences of symbols can be implemented in computers. By the same token, also the brain can implement syntactical relations. Both brains and computers are nothing but certain particle configurations. In order to get on this basis to thought, one has to answer the question of how the transition from syntax to semantics takes place, that is, how it comes about that certain regular sequences of symbols have a meaning.

John Searle (1980) famously argues that functionalism cannot answer this question: Searle imagines himself to be locked in a room in which he receives from the outside sentences written in Chinese. He has a rulebook at his disposal that tells him which sentences in Chinese to pick as output when he receives specific sentences as input. Searle thereby is in the position to give written replies in Chinese to the people outside of the room. However, he obviously does not understand what these sentences mean. He does not learn Chinese in that manner. By means of this example, Searle intends to show that functionalism about mental states cannot get beyond purely syntactical relations, which are insufficient for meaning.

The standard functionalist reply to Searle's challenge is to point out that it is not the person locked in the room that understands Chinese, but the entire system consisting of the person enclosed in the room, the rulebook and the interaction with the environment via inputs and outputs.[7] Hence, the idea is that syntax plus causal relations to the environment yield semantics. The syntax provides for the inferences that licence the transition from one sentence or thought to other sentences or thoughts, whereas the causal relations make up for the link with the world that then, together with the inferences, results in reference and meaning. The theory of Jerry Fodor (e.g. Fodor 1987, in particular the appendix) is the most prominent example of such a view. This is a completely naturalistic account that locates meaning in the ontology of the scientific image of the world. The availability of such an account is the reason why functionalism is widely accepted when it comes to thought. The main problem for functionalism is considered to be the qualitative aspect of consciousness.

Both sides in this debate tend to neglect an important feature of thought and rationality. There is more to the rational side of the mind than syntax plus causal relations to the environment. There also is normativity. Normativity evidently comes into play when actions are concerned. However, persons deliberate not only about what they should do, but also about what they should think. Both beliefs and actions are subject to providing a justification on request. Otherwise, they would not be beliefs and actions.

It is obvious that there is a serious problem to locate normativity in the ontology of the scientific image of the world. That image is concerned with facts. However, one cannot draw normative conclusions from statements about facts, on pain of committing what is known as the naturalistic fallacy. Thus, to take a common sense example, if science establishes that smoking is unhealthy, it does not follow from this that one *should* not smoke. This can follow only given a normative premise, such as the one that one *should* not do anything that is unhealthy, *ceteris paribus*. Hence, statements about facts cannot entail normative statements, unless a normative premise is added. However, it seems that there are no normative statements included in the statements that describe the primitive ontology of the scientific image, whatever that ontology may be. Let us therefore note the issue of *how to deal with normativity as the other big*

---

[7] See notably Dennett (1991, ch. 6), and the discussion between Searle and Dennett in Searle (1997).

problem for the scientific image as complete image of the world, besides the issue of sensory qualities and conscious experience.

The problem is this one: on the one hand, it seems impossible to locate norms within the scientific image, since propositions about facts cannot entail propositions about what persons should do on pain of committing the naturalistic fallacy. On the other hand, normativity cannot be eliminated as, for instance, witches can be eliminated: there is no need to locate witches in the scientific image. They do not exist. There are only erroneous beliefs of persons about witches. However, normativity is indispensable, since we have to act and thus to take decisions, that is, to make up our minds about what we should do.

One may envisage drawing on biological functionalism to resolve this problem. Consider a primitive ontology of science that entails the statement that organisms seek to optimize their fitness, that is, seek to optimize their opportunities for survival and reproduction. This ontology can certainly explain a good amount of the behaviour of organisms in terms of pursuing the optimization of their fitness in a given environment. Also many actions of humans can be explained in that manner. However, these are all explanations in retrospect. It does not follow from the statement that organisms seek to optimize their fitness that humans *should* do whatever they estimate to be optimal for their survival and reproduction. Furthermore, it does certainly not follow that they *should* employ whatever means are appropriate to enhance their survival and reproduction in a given situation.

If beings have thought and thus are persons, they can think or deliberate about what to do. They thereby go beyond biological needs and desires. They can do otherwise than simply follow their biological inclinations. If they can do otherwise than simply having their behaviour determined by their biological needs and desires, they are free in their actions. If they are free, they can be asked to justify their actions. By contrast, it makes no sense to ask for a justification of the behaviour of animals that simply follow their biological inclinations—such as cows, wolves, cats and dogs—, although they have conscious experience. From the freedom in action then follows that biological facts *per se* cannot be a justification for the decisions that persons take. The issue is which decisions persons should take, given the biological and all the other facts.

A similar remark applies to employing common sense functionalism in this respect. Starting from the normative, moral attitudes that people *de facto* have in a society at a certain time, one can formulate functional definitions of these attitudes that finally come down to definitions in terms of dispositions for behaviour—that is, what people do under certain circumstances given their attitudes. One thereby locates these attitudes in the ontology of the scientific image. For instance, Jackson (1998, chs. 5–6) advocates such a functionalism. However, again, if beings have thought and thus are persons, they can think or deliberate about what to do. They can therefore be asked to justify their actions. But then, again, the fact of having certain attitudes is not *per se* a justification. The deliberation is precisely about whether or not the initial attitudes are *right* or *correct*. Again, the fact of having certain attitudes cannot imply that these are the *right* attitudes when it comes to the question of what one *should* do.

It becomes now evident why the naturalistic fallacy is a fallacy: when it comes to norms, it comes to justifications, to giving and asking for reasons. This presupposes the freedom to deliberate about what to do and what not to do. That is how norms come into play and with them justifications. And that is why the naturalistic fallacy is a fallacy: it omits the normative premise that addresses the deliberations of persons—what the person *should* do. Hence, if a being is a person, it is responsible in a literal sense for what it does. One can ask it to give reasons and thereby to justify what it does. If a being does not respond to such requests, it is not a person. This statement has to be qualified by also including beings that develop into persons under normal circumstances, such as human infants: they respond to treating them as persons by developing (gradually) into persons. Moreover, a being can still be a person however little biological similarity it may bear to a human. It is only that under normal circumstances, humans are persons, and as far as we know as yet about the universe, only humans are persons.

Normativity concerns not only action and its consequences for society, law, the state, etc., but also thought and thus the transition from syntax to semantics. Consider how John McDowell describes what it would take for a wolf to entertain beliefs:

A rational wolf would be able to let his mind roam over possibilities of behaviour other than what comes naturally to wolves. ... [This] reflects a deep connection between reason and freedom: we cannot make sense of a creature's acquiring reason unless it has genuinely alternative possibilities of action, over which its thought can play. ... An ability to conceptualize the world must include the ability to conceptualize the thinker's own place in the world; and to find the latter ability intelligible, we need to make room not only for conceptual states that aim to represent how the world anyway is, but also for conceptual states that issue in interventions directed towards making the world conform to their content. A possessor of *logos* cannot be just a knower, but must be an agent too; and we cannot make sense of *logos* as manifesting itself in agency without seeing it as selecting between options ... This is to represent freedom of action as inextricably connected with a freedom that is essential to conceptual thought. (McDowell 1995, § 3)

Hence, a person has to make up her mind not only as far as her actions are concerned, but also as far as her beliefs are concerned, and be it beliefs about simple everyday matters of fact. The reason is that beliefs involve concepts, and concepts do not simply follow from sensory impressions. As Kant points out, given the sensory input from the world, we are free as to what to believe:

If an appearance is given to us, we are still completely free as to how we want to judge things from it.[8]

Accordingly, Kant regards the concept of freedom "as the *keystone* of the whole structure of a system of pure reason" in the preface to the *Critique of practical reason* (quoted from Kant 1996, p. 139).

We can further illustrate this issue by drawing on what Sellars (1956) denounces as the "myth of the given": this is the idea that something that is simply given to the mind has as such an epistemic status in being in the position to justify beliefs and actions. According to Sellars, thus, for instance, sense impressions, construed as the effects of interactions of a person with her environment, cannot, qua being the result of physical

---

[8] *Prolegomena* § 13, note III; quoted from Kant (2002, p. 85); "Wenn uns Erscheinung gegeben ist, so sind wir noch ganz frei, wie wir die Sache daraus beurteilen wollen" in the German original.

causal processes, *justify* anything. By the same token, supposedly innate ideas cannot as such justify anything. The reason is that with respect to whatever is given to her mind, a person has to take the attitude of endorsing what is given as a reliable source of knowledge and guide for actions; only thereby she confers to it an epistemic status. Hence, the person has to decide herself in deliberating about what is given to her which beliefs she *should* adopt and which actions she *should* take.

Consider also what Descartes says in the third *Meditation* about the idea of God: the fact that this idea is given to him does not imply that he *should* believe that there is a God. Only *his* deliberation about this idea, *his* examination of it leads to that conclusion. Consequently, any belief, too, is subject to a justification—although for many beliefs as for many actions, there usually arises no demand of justification. Nonetheless, when there is a demand of justification, sensory inputs cannot *per se* justify the belief. The issue then is whether the sensory inputs are in the situation at hand a reliable source for forming the belief in question. This can only be settled by invoking other beliefs. The issue then is in the last resort which network or system of beliefs accommodates best or is most coherent overall with respect to the available evidence. One can therefore justify any one belief only against the background of many other beliefs that are not called into question. Hence, one cannot call into question all beliefs at once, for there would then be nothing left that can serve as justification. Nevertheless, for any one belief considered individually, one can call this belief into question and thus demand a justification for it.

It is certainly true that action has priority over belief in the sense that when one deliberates, the first question is "What should I do?" and not "What should I believe?". That is, the question of what to believe arises in the first place with respect to action. This truth has been stressed by pragmatism from the American pragmatists of the nineteenth century on to pragmatism in today's philosophy of mind. Notably Martin Heidegger turned this issue into an anti-Cartesian and anti-representationalist epistemology and anthropology in *Being and time* (1927), followed by Richard Rorty (1980) and others. Most recently, Robert Brandom (2015, in particular ch. 1, part II) seeks to undermine Sellars's scientific realism on that basis.

That notwithstanding, science is the attempt to understand the world as it is, that is, to form beliefs about the world by adopting a point of view from nowhere and nowhen and thereby separate facts from norms. Nothing in what pragmatism drives home with respect to the priority of action infringes upon the possibility of persons to engage in what Aristotle describes as the development of *theoria* at the beginning of the *Metaphysics*, and nothing infringes upon the validity of that activity.

Thus, also the above mentioned functionalist account of meaning in terms of causal roles lays stress on beliefs about the physical world having the purpose to represent what there is in the world. Nonetheless, representation, insofar as it is situated at the level of beliefs and thus employs concepts, cannot simply be a matter of syntax plus causal relations to the environment. The reason is the difference between regularities and rules. There are regularities of particle motion. But particle motion just happens. It makes no sense to ask for justification here.

By contrast, if a person forms a belief—and be it such a simple belief as "This is blue"—, she employs at least one concept. She thereby follows a rule that fixes what is correct and what is incorrect in applying the concept. More precisely, she does so only if she is aware of her employing a concept being subject to a differentiation between correct and incorrect. This is what distinguishes rule-following from regularities, and this the reason why beliefs are subject to a justification. Rule-following as necessary and sufficient condition for mastering concepts has been worked out notably by Ludwig Wittgenstein in the *Philosophical Investigations* (1953, §§ 138–242) and the interpretation of Wittgenstein by Saul Kripke (1982). Based on rule-following, a normative, inferentialist semantics has then been elaborated on by Wilfrid Sellars (1956), Donald Davidson (1984, essays 9–12) and Robert Brandom (1994) among others.

Sellars (1956, § 14) illustrates the distinction between regularities of motion and following of rules in employing concepts by means of the example of a clerk in a necktie shop that becomes equipped with electric light in the early 1950s. The clerk thus sees for the first time that electric light changes the colours in which objects appear to persons. He continues to employ the colour concepts according to the manner in which the objects appear. He thus recommends to a client in the shop a handsome green necktie that, upon examination in daylight, turns out to be blue.

That is, it appears as blue in daylight and as green in electric light. Nonetheless, it is blue independently of the light conditions in which it is observed.

The point of this example is to illustrate in the first place semantic holism: the meaning of any concept, even concepts close to sensory experience such as "blue" and "green", does not consist in a relationship to sensory experience or a causal relationship to the environment. It consists in the inferences to other concepts—in this case, concepts about the standard conditions for judging the colours of objects and what defines these conditions. The standard conditions for judging the colours of objects include daylight. The concepts are employed in such a way that objects are not regarded as changing their colour when moved from daylight to electric light, although the colour in which they appear may thereby change.

Furthermore, this example illustrates the social and normative character of meaning: the inferences that determine meaning are fixed by social interactions, namely by what is fixed as the correct or incorrect use of concepts in the interactions in a community. Thus, one can imagine a community that applies colour concepts according to the way in which things appear. The issue here is not what (if any in this case) is the correct theory about the world, but how we acquire concepts, given that sense impressions cannot impose concepts upon us. There hence is no semantics without pragmatics. Pragmatics determines the use of concepts and thereby determines their meaning by fixing the rules that fix conceptual content.

Brandom (1994, part one) spells this view out in terms of meaning being constituted by normative practices of commitment, entitlement and precluded entitlement. For instance, if under appropriate circumstances, a person utters the statement "The animal over there in the water is a whale", she thereby is committed to statements such as "The animal over there in the water is a mammal", she is entitled to statements such as "The animal over there in the water is huge" and she is precluded from being entitled to statements such as "The animal over there in the water is a fish". The meaning of the concept "whale" thereby consists in the inferences that its use licences according to the norms of commitment, entitlement and precluded entitlement that are endorsed in a community.

These norms are of course subject to change; the normative practices that fix meaning are in continuous evolution.

These normative practices are irreducible to regularities of behaviour. The crucial point is again the one that Kant makes in saying that when sensory impressions are given to a person, the person is completely free how to judge things, that is, how to form concepts and beliefs (Prolegomena § 13, note III). This freedom consists in the person having normative attitudes by deliberating about what she should believe. On this basis, she forms concepts and beliefs in engaging in social, normative practices that are the expression of these normative attitudes and that determine conceptual content in terms of commitments, entitlements and precluded entitlements.[9]

This latter point brings in social holism, that is, the stance according to which it is constitutive for someone to be a person to participate in social practices of a mutual attribution of beliefs, of giving and asking for reasons. Social holism in this sense is endorsed by Wittgenstein (1953, §§ 138–242), Sellars (1956), Davidson (1984, essays 9–12) and Brandom (1994) among others. The argument is this one: only social, normative practices can provide a person with a distinction between following a rule correctly and failing to do so. For a person taken in isolation, everything that she does would count as correct for her, given that for everything that she does, there is a rule possible according to which what she does is correct (see in particular Kripke 1982 for details).

However, one can voice the following reservation against this distinction between rule-following, which is normative, and regularities of behaviour: by the same token as for anything that a person does, a rule can be conceived in retrospect such that everything that the person does counts as following that rule, so for any particle motion that happens in the universe, a regularity can be formulated in retrospect under which the particle motion in question comes. Therefore, in a certain logical sense, also determinism is trivial. Given all the particle motions in the universe, one can always formulate a logical system of regularities that is deterministic: the propositions stating these regularities and the propositions describing an initial state of the universe entail the entire set of propositions

---

[9] See e.g. Esfeld (2001, ch. 3) for a detailed account.

that describe the evolution of the particle motion in the universe. But this does not imply that there are laws of nature for the simple reason that this logical trivial sense in which determinism can always be obtained does not imply that the statement of such a logical system of regularities and an initial state is shorter or in any sense simpler than a long list that merely registers all the particle motion that happens in the universe. If laws of nature can be formulated, there are salient patterns or regularities in the particle motion in the sense that the same motion occurs in sub-configurations of the universe again and again. These patterns or regularities are real in the sense that they are independent of observes and their languages.

By the same token, social practices do not create free floating rules. Again, the logical possibility of the formulation of any odd rule is to be distinguished from the real rules that seek for objectivity in that they seek to capture the patterns or regularities that there are in the world independently of any language. Thus, it is correct to regard whales as mammals and incorrect to take them to be fish, because the salient regularity in the world is the way of reproduction and not the environment in which an organism lives. In general, what is correct and incorrect in the use of language is in the first place determined by the world, that is, the salient regularities that there are in the changes that occur in the world; but we can find out what is correct only in the mentioned social practices.

This objectivity notwithstanding, the gap between regularities and rules remains. The rules have to be formulated by us. Thus, a community may employ the concept "whale" in such a way that employing this concept entails the commitment to regard whales as fish. The community then does not seize some of the salient regularities in the world. Hence, even if there may be exactly one best system to represent what there is in the world, the conceptualization of the best system goes via the evidence that we have at our disposal through sensory impressions, and our sensory impressions cannot fix our concepts. With the freedom in forming concepts comes in rule-following and normativity. The logical sphere of giving and asking for reasons is closed by justifications, which cannot reach out into the physical realm, on pain of what Sellars (1956) denounces as the "myth of the given". The justifications can invoke the regularities in the physical realm, but these regularities cannot in themselves be a justification for anything. This leads us to a coherentist theory of justification, as already mentioned above: one cannot invoke anything outside the sphere of beliefs as justification for a belief.

Nonetheless, justification being coherentist does not touch upon the architecture proposed in this book, namely, the formulation of a primitive ontology whose propositions entail all the other propositions via the explained procedure of location, at least as far as all the propositions about the physical domain are concerned. The point is that the justification for the chosen primitive ontology is coherentist: it has to yield the overall best explanation. Hence, the primitive ontology is not a foundation of knowledge, although it entails all the other propositions given functional definitions (at least as far as the scientific image of the world is concerned).

3.3   The Scientific and the Manifest Image

Sellars (1962) contrasts the scientific image of the world with what he calls the manifest image. The latter is the image that is based on our sensory experience of the world and the conception of ourselves as persons, that is, as thinking and acting beings in the world. However, the contrast is not between science and common sense. Common sense is pre-scientific and pre-philosophical. It leaves open whether or not the common sense experience or conception of the world and of ourselves admits of a scientific account. The manifest image is a *philosophical* theory of the world that is centred on persons. It endorses persons and their characteristic features as ontological primitives.

Both the scientific and the manifest image have their origin in the common sense experience and conceptualization of the world and ourselves as persons in the world. The scientific image then poses theoretical entities that are characterized only by a few basic physical features, such as point particles individuated by distance relations and the change in these relations. It seeks to locate everything else in these basic physical features through the mentioned functionalist procedure. Progress in science is progress in that location. The manifest image, by contrast, rejects these ontological primitives and, consequently, is opposed to the mentioned functionalist procedure. Its proponents can play the card of this procedure leaving something out, as claimed by the explanatory gap argument discussed at the end of Sect. 3.1. But this is not the argument for the manifest image. The explanatory gap argument does not have the force to establish the conclusion that persons are ontologically primitive: one can reject the intuition that the functionalist method breaks down when it comes to persons and their consciousness.

The argument to that conclusion is this one: persons have to take decisions and thus to answer the question of what they should do, including which beliefs and theories they should adopt. This is what is in any case correct in the Cartesian argument that one cannot doubt that one thinks. Consequently, normativity is presupposed for the very formulation of the scientific image. The scientific image depends on thought for its existence as *image*, that is, as theory that employs concepts whose meaning is determined within normative practices of giving and asking for reasons. Formulating and endorsing the scientific image is a choice that persons make and that can only be justified within the sphere of normative attitudes of giving and asking for reasons. The referents of the theory—whatever the theory poses as existing in the world—cannot impose the acceptance of the theory on persons. In that sense—as the beings that formulate and justify theories in normative practices of giving and asking for reasons—, persons are indispensable and thus primitive: whatever the theory is, persons have to conceive, endorse and justify the theory in question.

Thus, suppose that the theory says that everything that exists is matter in motion. Nonetheless, one cannot claim that the matter in motion in the world imposes this theory on us, because the theory itself is nothing but a configuration of the matter in motion in the sense that it is nothing beyond the thoughts that persons have, and these are identical with certain particle motions in their brains. However, any such claim is itself conceived, endorsed and justified in the normative web of giving and asking for reasons. As Jaap van Brakel (1996, p. 280) puts it, scientific categories derive their existence in part from normative, methodological criteria grounded in the manifest image. As pointed out in the preceding section, whatever sensory input we get from the world, we are free what to make of it in the sense that we are free to deliberate what rules to set up both for thought (concepts) and action. This freedom and the normativity that comes with it cannot be located in the scientific image and its primitive ontology.[10] This, then, is the argument for persons being ontologically primitive on which the manifest image bases itself.

Consequently, the view of ourselves as expressed in the manifest image is immune to whatever progress neuroscience may realize in the future. It does not make sense to claim that the view of ourselves as thinking and acting beings that have to make up their minds as to both thought and action through normative attitudes may be proven wrong by future progress in neuroscience. In this vein, for instance, Al Mele (2014) develops an extensive argument why much discussed experiments in neuroscience and psychology—such as the ones presented in Benjamin Libet (2004)—do not refute the claim that we have free will. Of course, such experiments and neuroscientific progress in general may contribute to a better understanding of ourselves. It may, for instance, reveal that we seek to find reasons in retrospect for actions that were merely impulsive or emotional reactions to certain stimuli. But, still, we could have done otherwise, if we are persons. That is, we should receive such research results as an appeal to do better in the future, that is, to try to live up to the potential that comes with being a person by reflecting upon what we should think and do. Furthermore, such future progress may also show which brain damages destroy the physical basis for being able to do so.

However, no such progress could prove the view of ourselves as persons as conceptualized in the manifest image to be an illusion. The very claim of that view being an illusion would be a manifestation of the validity of that view. Again, this claim would be conceived, endorsed and justified in normative practices of giving and asking for reasons. For it to have meaning, for there being reasons to take it to be true, etc., it would have to be situated within these practices. That notwithstanding, it may turn out that we evolve in the future in such a way that we lose the status of being persons and return to the state of brutes, abandoning thought and action. However, this would then not be the result of a discovery of neuroscience or psychology, but the consequence of the abandonment of science.

---

[10] Cf. also the fundamentalism about reasons advocated by Scanlon (2014).

Since the manifest image can thus not be dismissed out of hand, the basic issue obviously is how to bring these two images together. However, the problem is that both these images claim to be complete; but they cannot be both true and complete. The completeness claim of the manifest image is this one: starting from persons as ontologically primitive, the method is the one of conceiving everything else in analogy to persons with there being different degrees of similarity to persons—for instance, cats are more similar to persons than trees, etc., and the similarity to persons fades out in inorganic matter, although it never disappears completely. That notwithstanding, one may recognize a primitive material stuff that has no form and thus nothing analogous to persons, as in ancient philosophy, notably both in Plato and in Aristotle. Nevertheless, each object then consists of a form (that is, something analogous to persons) or the participation in a form and primitive stuff. Thus, one pursues a top down methodology in contrast to the bottom up methodology of the scientific image. Scientific theories consequently describe the universe by abstracting from the features that are analogous to persons. They can be true, but the truth that they discover is only a partial truth also as regards the physical objects. Science fails to reveal the essence of these objects, which consists in features that are analogous to features that characterize persons.

Hence, there are two complete images of the world that contradict one another. On the scientific image, matter in motion is ontologically primitive. Everything else—including human beings and their minds—is introduced in terms of its causal or functional role for the motion of matter. On the manifest image, persons are ontologically primitive. Everything else—including inorganic matter—is conceived in analogy to persons. The basic issue then is to integrate the elements on which the one image lays stress into the other image. If this could be done successfully, the other one would be eliminated as an image of the world. If everything is analogous to persons, we need the scientific theories as useful instruments for predictions; but we do not need the scientific image. If everything is matter in motion, we need the language that is centred on persons as useful instrument of communication; but we do not need the manifest image.

Thus, if one starts from the scientific image, the issue is to integrate everything, including what characterizes persons, into that image through the progress of science and thereby to make the manifest image obsolete. That is, what is made obsolete is not common sense and our experience of the world and ourselves, but the conceptualization of this experience in an image of the world that is centred on persons. By the same token, if one starts from the manifest image, the issue is to integrate what science tells us about the world into an image of the world that endorses persons as primitive. One does not abandon the view that there is matter in motion; but one rejects the claim that one can understand the world on the basis of matter in motion. One thereby makes the scientific image obsolete.

Science cannot settle this dispute, because science is not identical with the scientific image of the world. One can endorse science including its reductionist methodology and refrain from applying it to persons. The scientific image is the *philosophical* view that subscribes to a completeness claim to the effect that science discovers *all* there is in the world. There is not anything more to what there is in the world than what is conceptualized in a scientific theory. By the same token, the manifest image is the *philosophical* view that lifts the status of common sense as it is centred on persons to a completeness claim to the effect that *everything* is analogous to persons. One can therefore consider the clash between the scientific and the manifest image as the great fight about being, to use Plato's terms in the *Sophist* (246a).

If one adopts the view that the *scientific image is the complete image of the world*, it is no problem to apply the procedure of functional definitions in terms of a causal role for matter in motion to everything. One can do so just by stipulation, that is, by definition, as pointed out in comment (2) in the discussion of functional definitions in Sect. 2.1. Furthermore, there is no principled obstacle to formulating an explanation of everything, including the beliefs and actions of human beings, in terms of its functional role for, in the last resort, particle motion. In the case of human beliefs and actions, this in the first place a biological functional explanation in terms of enhancing fitness. But enhanced fitness is in the last resort a form of particle motion.

However, these are explanations in retrospective. In retrospective, one can in principle always formulate an explanation of everything in terms of its functional role for matter in motion. Recall that first comes the motion of the objects, then come the laws and with them explanations. More precisely, in order for such explanations to be available and for them to be applicable to human thoughts and actions, one has in the first place to endorse the scientific image. Its endorsement and justification take place in the normative practices of giving and asking for reasons, which presuppose persons as ontologically primitive, because it is persons that formulate, endorse and justify theories and worldviews. Their referent cannot impose them on persons, whatever the content and referent of these theories and worldviews may be.

In other words, also a biological functionalist account of human thoughts and actions cannot touch upon the fact that in forming her beliefs and deliberating about her actions, a person has to navigate in the normative web of giving and asking for reasons. That navigation cannot be determined by what enters as the primitive ontology into the scientific image, because the person first has to make up her mind about what she should believe with respect to the world. A similar remark to the one made above when mentioning alleged neurobiological and psychological evidence against free will applies here: explanations of thought and action in terms of biological functions are best received as an appeal to do better in the future, that is, to live up to the potential that comes with being a person and weigh cogent reasons for one's thoughts and actions.

When one situates oneself in the normative web of giving and asking for reasons, one is thereby committed to persons as ontological primitives. This opens up the possibility that they are the primitive ontology. That is to say: as one can try out the stance that the scientific image is the complete image of the world, so one can try out the stance that regards the *manifest image as the complete image of the world.*

If persons are the primitive ontology, then it is not a credible method to seek to locate everything else in persons by means of introducing everything else in terms of a functional role for the deliberative activities of persons. With respect to the physical world, this would amount to rejecting the existence of nature, recognizing only sensory impressions in the minds of persons. Nonetheless, functionalism is a powerful tool also in the context of the manifest image. But the issue then is a *normative functionalism* that is limited to what concerns persons as thinking and acting beings. Thus, on an inferential role semantics combined with a normative pragmatics as proposed by Sellars (1956) and worked out by Brandom (1994), the meaning of a thought consists in the inferences to other thoughts as well as actions; these inferences are determined by social, normative practices, as mentioned in the preceding section. In this way, meanings are located in the social, normative practices that make persons the beings that they are. There then is no need to admit meanings as abstract objects (Platonic ideas). They are reduced to what persons take themselves to be committed to, to be entitled to and to be precluded from being entitled to.

This procedure can in principle be applied to all the candidates for abstract objects, including mathematical objects such as numbers. The resulting position then is a sort of nominalism, but a normative one. This procedure can also be applied to moral norms. One can locate them also in what persons take themselves to be committed to, to be entitled to and to be excluded from being entitled to as determined by the normative, social practices of a community. If one considers moral norms to transcend community agreement (such that a whole community can go wrong about the moral norms that it endorses), then one can seek to locate moral norms in what it is to be a person qua a being that is responsive to norms, including moral norms. To put it in the terms of Karl Popper (1980), in the stance that endorses the manifest image as the complete image of the world, one can employ functionalism to reduce Popper's world 3 (abstract objects) to his world 2 (persons). But one cannot employ functionalism to reduce world 1 (physical objects) to world 2 (persons). One can only seek to reduce world 2 (persons) to world 1 (physical objects) within the stance that regards the scientific image as complete.

When it comes to the relationship between persons and the physical world, the method employed in the stance that takes persons to be the primitive ontology is analogy instead of reduction. As mentioned above, everything in nature is conceived in analogy to persons. However, the problem with this method is that it does not provide the explanations that successful reductions achieve. By starting from As and saying that all Bs are analogous to As, one does not answer the question why there are Bs in the world. By contrast, one answers this question if one replies that there are Bs in the world because some specific configurations of As are such that they fulfil the functional role that defines *Bs*. Hence, starting from persons as the primitive ontology, we need answers to the questions of why not all that exists are persons, viz. why there are entities that are to different degrees analogous to persons. However, we do not get answers to these questions in the manifest image.

The manifest image has to accommodate scientific explanations, and be it only for their predictive power, because persons rely on this predictive power when they deliberate about their actions. But it cannot accommodate them as explanations of the objects to which they refer: their essence is not to be matter in motion, but to be in some way or other analogous to persons. Hence, they only have an instrumental value as information about the consequences that different choices of actions are likely to have. Nevertheless, in order to have such an instrumental value, they have to be true, that is, have to consist in true propositions; the consequences of our actions obtain independently of what persons believe. That notwithstanding, the scientific explanations do not reveal the essence of the objects in nature.

This can be illustrated in terms of the geometrization of the things in nature through a primitive ontology of point particles individuated by relative distances and their change. The laws formulated in the framework of this ontology provide information about the evolution of the relative positions of the objects, which is all that we need as orientation for our actions. But this ontology conceives the physical objects as featureless. From the point of view of the manifest image, this ontology thereby abstracts from all the qualitative features of physical objects— such as colours—, which are related to our sensory experience. From this point of view, it misses the essence of the things in nature. Of course, it still is a long way to go from qualitative features such as colours to conceiving an analogy to persons. Nonetheless, assume for the sake of the argument that one can spell such a way out. The point at issue then is that scientific explanations come in only as instruments for predictions that are useful for actions, but not as revealing what the things in nature are. Hence, the manifest image has to jettison scientific realism also with respect to the theories of physics.

Even if one leaves aside that the manifest image cannot match the explanations that one achieves in the scientific image through functional reductions, there is an obvious problem with the idea to conceive everything in analogy to persons. Sellars (1962) considers the example of a pink ice cube. The ice cube is pink through and through as far as the manifest image is concerned. Dividing it into parts does not remove its colour. One can take this as an illustration of what is meant by thinking in terms of analogies: the features in question will never fade away—that is, the features of personhood, or, in this case, colour. However, this is not true. If put under a microscope and if the resolution is high enough, one will see with the naked eye that the parts of the pink ice cube distinguished under the microscope are not pink themselves. There is nothing pink in them, not even a shade of pink.

Already when it comes to non-human animals, it is debatable to what extent they are analogous to persons. But some of them have at least sensory experience. However, when it comes to inorganic matter, it is widely implausible to maintain that, say, stones still are somehow analogous to persons, not to speak of microphysical particles. What could be the mental features in electrons, however rudimentary they may be? To put it differently, the manifest image is committed to some sort of panpsychism. But there is no reason to endorse panpsychism apart from the logic of pursuing a primitive ontology of persons within the framework of regarding the manifest image as the complete image of the world, or the logic of admitting phenomenal consciousness (qualia) as an ontological primitive in order to solve the hard problem of consciousness.[11] Furthermore, pursuing a dual aspect theory of a mental and a physical aspect in everything is of no help either. The question is how these two aspects hang together. Enlarging their extension such that these two aspects are present in everything does not contribute to answering that question (as is already evident from the remarks on emergence at the end of Sect. 3.1).

In sum, both with respect to the scientific and with respect to the manifest image one can set out an argument that comes pretty much close to a *reductio ad absurdum* of the claim that either one of these images is the complete image of the world. In the case of the manifest

[11] See the papers in Brüntrup and Jaskolla (2017) for contemporary research on panpsychism, as well as Benovsky (2019).

image, the *reductio* is the obligation to conceive everything, down to the microphysical particles, in analogy to persons. The *reductio* in the case of the scientific image is the impossibility to integrate persons insofar as they deliberate about rules for thought and action into this image; but the scientific image relies on persons thus deliberating, since this is the manner in which it is conceived, endorsed and justified. One may say that claiming that the scientific image is complete comes close to a performative contradiction: the content of the claim that everything is matter in motion contradicts its performance as *claim* that is situated in the normative web of giving and asking for reasons in which persons are primitive. Hence, we have two images each of which claims to be complete and true and each of which contradicts the other one; but it seems that neither of them can be the true and complete image of the world.

It may also be instructive to illustrate this situation in terms of Popper's (1980) three worlds. The upshot then is that one cannot do with only one of these worlds. Doing with only world 1 (the scientific image being complete) runs into an impasse because the very conceptualization of world 1 is carried out within world 2. Doing with only world 2 (the manifest image being complete) commits one to considering everything as analogous to persons. Doing with only world 3 eliminates every concrete entity, leaving us only with abstract objects. Endorsing all three worlds, as Popper (1980) advocates, again commits one to surplus structure: there is much more in world 3 then ever conceptualized by persons or instantiated in the physical domain (world 1). Furthermore, there is the mentioned functionalist procedure available to reduce to world 2 what is reified by Popper to a world 3. This leaves us with the possibility to combine in principle any two of these worlds. Endorsing world 1 and world 3 runs again into the problem that the very conceptualization of these worlds takes place in world 2. Hence, world 2 is indispensable. Going for world 2 and world 3 amounts to an idealism that replaces physical objects with sense impressions plus abstract objects. The upshot then is that, again, we have to find a way to bring world 1 (the scientific image) together with world 2 (the manifest image).

3.4   The Synoptic View

Sellars (1962) calls for a synoptic view that brings both images together as a whole, in his words a "stereoscopic vision, where two differing perspectives on a landscape are fused into one coherent experience" (1962, section I). This implies some sort of a dualism of both images, although each of these images claims to be complete. The obvious question for any dualism is why one should admit two primitives and cannot do with just one. However, in this case, there is a well-founded motivation for a dualism of both matter in motion and persons as ontological primitives. There is a *principled* argument why the ontology of science is not complete when it comes to persons. To integrate normativity into this ontology is not an issue of further progress in science, but in principle excluded, whatever progress (neuro)science may make. By the same token, there is a *principled* argument why the ontology that takes persons as primitive and conceives everything else in analogy to persons is not correct at least when it comes to inanimate matter.

The pressing issue for a dualism of matter in motion and persons hence is not how to motivate this dualism, but how these primitives are related with one another. Consider Cartesian dualism of *res extensa* (matter) and *res cogitans* (mind, persons). The problem for Descartes is that on the one hand, he has to conceive a point of contact of both these substances. However, on the other hand, any conceivable point of contact where the non-physical mind is supposed to get in touch with the material domain is widely implausible (such as a point of contact located somewhere in the brain, as in the pineal gland for Descartes). This is the real problem for interactionist dualism. The problem of causal interaction or causal intervention of something non-material in the material domain then is a follow up of this problem.

The crucial issue of a point of contact is not resolved by moving from interactionist dualism (as in Descartes) to psychophysical parallelism (as in Spinoza and Leibniz). It is not true that in the case of psychophysical parallelism, mind and matter is brought together only in God. Particular minds have to be associated with particular bodies. This requires a point of contact where the individual mind touches upon a particular body.

The fact that there is no convincing solution to this problem suggests that dualism cannot be a strong contender for the truth as long as it is conceived as a dualism of two types of entities of the same category, such as a dualism of a physical and a mental substance as in Descartes, a dualism of physical and mental properties from Spinoza and Leibniz on to contemporary non-reductionist positions, a dualism of physical and mental states of affairs, facts, aspects, etc. As long as one assumes that there are two different types of the same category, the problem of their relationship will not find a satisfactory solution. That is to say: a dualism of matter in motion and mind is well motivated, but not its elaboration in terms of two types of entities of the same category. There is no convincing argument why there should be two types of entities of the same category, as the problem of their point of contact makes evident.

The strongest argument against the scientific image being complete is the one according to which this image is itself conceived, endorsed and justified in normative attitudes of giving and asking for reasons. This argument gets us to endorsing persons as ontologically primitive. However, *in virtue of being characterized by normative attitudes, persons then are not an ontological primitive on the same level as, in addition to and thus of the same category as the ontological primitive of the scientific image, namely matter in motion.* If the mental is irreducible to the physical, this is so because norms are not facts in the world. They come into existence through persons coming into existence by deliberating about what they should believe and do, thereby creating a normative web of rights and obligations, of commitments, entitlements and precluded entitlements in which they situate themselves.

That notwithstanding, this view is committed to persons being ontologically primitive, although thinking and acting beings exist only at certain places and times in the universe so that there are certain physical conditions for their coming into being. Hence, on the one hand, persons are ontologically dependent on matter in motion: they can exist only if certain physical conditions are fulfilled. On the other hand, they are an ontological primitive over and above the physical ones, since their normative attitudes cannot be reduced to matter in motion. The concept of a person as it is used in this book is intended to bring out both these aspects, namely the embodiment of persons as well as the irreducibility of their normative attitudes.

Insofar as they are ontologically primitive, persons with their normative attitudes can indeed be conceived in the same way as matter in motion, namely in the way of ontic structural realism. If they are primitive, they cannot be divided with the division resulting in other primitives. They therefore better be construed as points. Furthermore, the arguments against a bare substratum apply to persons in the same way as to matter points. These considerations get us to the following view: both matter and persons are points that are structurally individuated through the relations in which they stand. Matter points are individuated by their position in a web of distance relations. Persons or mind points are individuated by their position in a normative web of rights and obligations, commitments, entitlements and precluded entitlements that concerns beliefs as well as actions. As all there is to the matter points are the distance relations in which they stand, so all there is to the mind points are the normative relations into which persons enter through adopting normative attitudes.[12]

Both the distance relations and the normative relations are in continuous change. The normative relations change through every move that a person makes in her thoughts and actions. As the continuous change in the distance relations provides for an intertemporal identity of the matter points through the trajectories that they thereby trace out, so the continuous change in the normative relations provides for an intertemporal identity of the persons qua mind points. In both cases, a development to a completely symmetrical configuration is excluded: no two persons ever have exactly the same commitments and entitlements—this is already excluded by their different spatial and temporal positions and the different relations of kinship—, as no two point particles ever stand in exactly the same distance relations to all the other point particles. Hence, qua ontologically primitive, both matter and persons are points that are at every time as well as in their temporal development structurally individuated through relations of a certain type.

---

[12] Cf. also the normative, relationalist view of reasons of Scanlon (2014).

The categorical difference between matter points and person or mind points lies in the difference in these relations: distances that exist as a matter of fact versus norms that come into being through certain configurations of matter in motion adopting to themselves and others the attitude of taking themselves and the others to be situated in a web of rights and obligations. In adopting such an attitude, certain particle configurations create themselves as persons: in doing so—and only in doing so—are they persons. This difference in the relations implies that the normative relations only exist as long as persons continue to exist by adopting these attitudes. By contrast, the distance relations exist as a matter of fact so that the change in them is unlimited and unconditional (at least as far as scientific research is concerned; we can leave open here whether there is a God that brings the whole configuration of matter into being and may annihilate it).

The crucial issue of the point of contact between the sphere of facts and the normative sphere then is resolved in this way: there are sufficient physical conditions for responsiveness to norms. The ability to engage in social, normative practices is located in and thus identical with the motions of certain particle configurations. As Michael Tomassello (2014) works out, one can formulate a biological explanation of this ability in terms of the enhancement of fitness that cooperation between humans provides. Nonetheless, once these practices come into being, the norms that are determined in them are not located in the sphere of facts. They are not further facts in the world. They exist, as the matter in motion exists; but they are accessible only from within participating in these practices and thereby contributing to shape them. There is no perspective from nowhere and nowhere available to access these practices.[13]

Consequently, one may say that persons emerge during the evolution of the configuration of matter of the universe. The notion of emergence is applicable in this context, because it is clear that persons are irreducible to matter in motion and call for a commitment to a new ontological primitive. That notwithstanding, employing the notion of emergence does not illuminate or explain anything. Quite to the contrary, it draws the attention away from the crucial issue that norms are not new facts.

---

[13] See also the argument of Bilgrami (2006, pp. 62–64).

The difference is not one in existence or truth conditions. Existence and truth are unequivocal. Either something exists or it does not exist. Either a proposition is true, or it is not true. The difference is one of accessibility: without contributing to shape them in the case of taking note of facts in contrast to accessing norms only by contributing to determine what they are in adopting the attitude of treating oneself and others as persons.

Let us illustrate this view by drawing on a few historical sources. The idea that persons are subjects and that subjects are categorically distinct from objects (matters of fact) is pursued in the modern theory of subjectivity from Immanuel Kant on. Kant expresses this idea in terms of "the starry heavens above me and the moral law within me".[14] This is a dualism of matter in motion and persons, whereby persons are characterized by their responsiveness to norms.

In contemporary philosophy, one can associate the anomalous monism proposed by Donald Davidson (1970) with such a dualism (although the position is called "monism"). Persons are integrated into the physical world in the sense that there is no realm of being of the mental in addition to the physical realm. Davidson talks in terms of mental events being identical with physical events. In other words, all events are physical, and some physical events are also mental events. This is the monism in Davidson's position.

That notwithstanding, the conceptualization of an event as mental is irreducible to its conceptualization as physical. That is why Davidson's monism is anomalous. There are no psychophysical laws. The reason is that mental events, insofar as they have a meaning, are determined by normative criteria that are accessible only from within participating in the practices that set up these criteria. Davidson endorses social holism: it is constitutive for someone to be a person to participate in social practices of mutual interpretation, that is, mutual attribution of beliefs by giving and asking for reasons. These practices are autonomous in the sense that they take place according to criteria that cannot be reduced to

---

[14] Quoted from the translation Kant (1996, p. 269); "Der bestirnte Himmel über mir, und das moralische Gesetz in mir" in the German original, *Critique of practical reason* (1788), conclusion (Beschluß).

the parameters that figure in physical laws, as set out notably in Davidson (1984, essays 9–12).

This is not to deny that there are psychophysical generalizations, that is, regularities that explain in retrospective, as well as predict to a certain extent, the behaviour of persons by invoking their beliefs and intentions to actions. But these are not laws, since they are not exceptionless—or, to put it in less rigid terms, they have too many exceptions and are not counterfactually robust. They hold only in a rough and ready way against the background of stable social practices. In short, again, the idea is that the normative practices of giving and asking for reasons are autonomous and that when one situates oneself in these practices, persons are irreducible. Nevertheless, they are not facts (or events) in addition to the physical ones.

Davidson (1970) is the first to apply the notion of supervenience to the relationship between the physical and the mental. However, from today's point of view, the application of this notion is rather unfortunate. In contrast to notions such as identity and location versus non-identity and new ontological primitives, the notion of supervenience does not clarify this relationship. On the one hand, supervenience is not immune to reduction, as is clear at the latest since the widely disseminated arguments of Jaegwon Kim (1998): supervenience implies that there are sufficient physical conditions for everything mental. Consequently, the propositions that provide a true and complete description of the physical domain entail all the true propositions about the mental. On the other hand, the notion of supervenience yields no explanation. By contrast, the identity theory provides an explanation, if it is worked out in terms of functional roles of the mental that are realized by configurations of matter in motion. Hence, the theory of psychophysical supervenience buys into what one may see as the drawbacks of reductionism, without living up to what is the strong side of reductionism, namely its explanatory force.

A similar comment applies to the contemporary replacement of "supervenience" with "grounding"[15]: grounding, if applied to the mind-body problem, tells us that there are sufficient physical conditions for everything mental. Consequently, the propositions about the physical entail

---

[15] See in particular the papers in Correia and Schnieder (2012).

the propositions about the mental. However, these sufficient conditions and this entailment do not yield the benefit of a physical explanation of the mental: the concept of grounding expresses a correlation between the physical and the mental and accords ontological priority to the physical, but it does not explain this correlation and this priority.

Coming back to Davidson's position, there are sufficient physical (or physiological) conditions for beings to have the capacity to participate in social practices of mutual interpretation, giving and asking for reasons. One can put this in terms of sufficient physical conditions for responsiveness to norms. Humans usually fulfil these conditions. Other living beings such as wolves, bats and plants do not do so. Consequently, as far as humans are concerned, one can demand a justification for their behaviour, including their linguistic behaviour; but it makes no sense to do so when it comes to wolves, bats or plants. The issue only is sufficient, but never necessary physical conditions: Martians would also have the ability to engage in social practices, although they are not biologically similar to humans. Nonetheless, they would thereby also have the ability to participate in *our* social practices, since there is no principled limit to the possibility of translation. If something is a thinking being, that being is in principle in the position to participate in any social, normative practices and contribute to shaping them, as argued by Davidson (1984, essay 13).

However, it would be contrary to Davidson's position of an anomalous monism to assume that there are sufficient physical conditions for the rules that persons follow—that is, sufficient physical conditions for the meaning of thoughts and in general the content of the norms of thought and action that are determined in these practices. In this case, there would be psychophysical laws, since the propositions about the meaning and in general the norms would then be entailed by the propositions that describe the physical domain. Hence, the conclusion that imposes itself is that psychophysical supervenience fails for what is determined within the social practices of giving and asking for reasons as their content. That is to say: there are sufficient physical or physiological conditions for the ability to participate in social practices that determine meaning and norms in general. This ability supervenes on matter in motion; more precisely, it is some particular matter in motion. But once this ability in people that share a physical environment and thus interact locally is transformed into these people spontaneously adopting normative attitudes, thereby setting up social practices of giving and asking for reasons and thus creating themselves as persons, these practices are free. There are no sufficient physical (or physiological) conditions for what are the norms that are determined in these practices. In other words, there is no objective, scientific matter of fact what these norms are.[16]

These norms are not accessible from a third person perspective, that is, the point of view from nowhere and nowhen that characterizes science. They are accessible only from within participating in the practices of giving and asking for reasons and thereby contributing to shaping these norms. This restriction would also apply to an omniscient being (God). Such a being would know all the facts about the world. Nonetheless, also such a being would have to participate in these practices to know the norms that are determined in them; more precisely, by participating, it would contribute to determining these norms. This is a consequence of these norms not being accessible from outside the practices that determine them. This feature marks the difference between facts and norms: facts simply exist independently of anyone taking notice of them. Norms only exist insofar as there are normative attitudes of persons and are only accessible from within participating in the social practices that result from these attitudes.

This confirms again that we face the problem of how to bring the scientific and the manifest image together in a synoptic view not because our perspective or our knowledge is somehow limited. We can formulate scientific theories that apply to the universe as a whole from a perspective of nowhere and nowhen. Cosmology does so since antiquity. These theories (or some successors of them) may very well be true. They may even pose a principled limit on our accessibility of initial conditions such as the Heisenberg uncertainty relations in quantum theory (see Sect. 1.7). But such a limit is no axiom of a scientific theory. It follows from the—complete—description of the universe as a whole from a perspective from nowhere and nowhen that the theory in question provides.[17]

---

[16] See Lance and O'Leary-Hawthorne (1997) for a treatise on the implications of this for translations.

[17] Cf. for instance how Dürr et al. (2013, ch. 2) derive the Heisenberg uncertainty relations from the axioms of Bohmian quantum mechanics.

The issue is that any theory, including a theory of the universe as a whole construed from the point of view from nowhere and nowhen, can be formulated only from within participating in social, normative practices that determine its content. There is no other possibility for a theory or a whole image of the world, whatever its content may be, to be conceived, endorsed and justified. Davidson (1991, p. 164) expresses this crucial point in these terms:

> A community of minds is the basis of knowledge; it provides the measure of all things. It makes no sense to question the adequacy of this measure, or to seek a more ultimate standard.

Nonetheless, the community of minds as measure does not commit us to a relativism as regards truth: the community seeks to find out the truth about the world. The point only is that whatever is taken to be this truth has to be conceived and justified within the normative practices of a community that determine the content of any theory. There is no other measure of conceptualization and justification—and thus for judging candidates for the truth—available.

There are salient regularities in the motion of the matter in the world that carve nature at its joints and that yield natural kinds. Any science seeks to discover natural kinds that there are in the world as a matter of fact. Nothing is a science that does not consider natural kinds that result from salient patterns in the behaviour of the objects in its domain. Thus, the types of elementary particles in today's standard model of elementary particle physics are natural kinds in fundamental physics. The atoms in the sense of the elements of the periodic table of elements are natural kinds in chemistry. Bats, whales, lions and antelopes are biological species, etc. Therefore, for instance, if the content of the concept of a whale as used by a given community includes the commitment to the proposition "Whales are fish", this is objectively wrong, even if the community at a certain time was not in the position to know this. Given normative practices of determining conceptual content and setting up theories, the world then determines whether or not these theories are true. This objectivity is notably emphasized by Brandom (1994, ch. 8).

Defending the view that natural kinds are discovered through the mentioned social practices is relatively easy going in the case of thoughts, because it does not require more than scientific realism with respect to the physical domain. It is much less obvious how to defend a moral realism, because there is no domain of moral facts beyond the domain of physical facts that we seek to grasp by following rules for thought and action. If there is a truth about moral norms beyond community agreement, it has to come with the commitment to persons as ontologically primitive. The idea then is that by fulfilling the sufficient conditions for participating in normative practices and when doing so, persons have some fundamental rights and obligations in the light of which the moral practices of a whole community can be wrong, even if communities at certain times may not be in the epistemic position to acknowledge these fundamental rights and obligations. Thus, for instance, it may be that torture is always morally wrong, even if it was a widespread practice in the Middle Ages during the times of inquisition and even if whole communities (despite considering themselves to be Christian) may not have been in the epistemic position to realize that torture is wrong.

One stock objection that seems to hit also a dualism of normative attitudes and matters of (physical) fact is that intentions for actions, insofar as they are normative and subject to giving and asking for reasons, cannot have effects on the behaviour of humans. The reason is the completeness of the physical domain, which is also known as causal exclusion, as worked out by Kim (1998). For any physical event including the movements of persons, there are sufficient physical causes, insofar as there are causes at all. By the same token, any physical event including the movements of persons comes under natural laws in which only physical parameters figure, insofar as it comes under laws at all. However, it is just the point of Davidson's anomalous monism not to deny the latter issue. In general, there is a fallacy in this argument, which is of the same type as the fallacy that infers from determinism in physics that we do not have free will.

As argued in Sects. 2.2, 2.3 and 2.4, first comes the motion of matter including the behaviour of persons, then come the laws and with them causal relations and explanations, insofar as these are conceived within the scientific image of the world. Thus, the laws—and, as argued in Sect.

2.4, in particular the values of the parameters that enter into the initial conditions for the laws—, depend also on the behaviour of the persons in the universe. Of course, what is relevant for fixing the values of these parameters only is the behaviour of persons in the sense of their bodily motions. Nonetheless, some of these motions can be the expression of intentional states of the persons that are subject to giving and asking for reasons. Counterfactual propositions of the type "If a person had adopted other normative attitudes, her behaviour would have been different so that some particle motions would have been different" are true with respect to the real world.

The Super-Humean stance according to which the initial values of dynamical parameters in the initial state of the universe are fixed only by the motions that actually occur in the universe opens up also the possibility to recognize persons in the sense of beings that have normative attitudes through which they form thoughts and intentions to act as ontologically primitive while respecting mental causation without entering into conflict with physical causation. As explained at the end of Sect. 2.4, for this reason there is no need for specific dynamical parameters for free will that finally would have to be included in the laws of physics: the motions that are the expression of free will count among the motions in which these dynamical parameters are located and on the basis of which their initial values are fixed. Hence, the initial values that these dynamical parameters take are different in a universe with exclusively inorganic matter from what they are in a universe in which some complex physical systems are organisms and a universe in which some complex physical systems are humans with free will.

It seems that the three stances sketched out in this and the preceding section exhaust the logical space: either the scientific image is complete, or the manifest image is complete, or a dualism that somehow combines the ontological primitives of both these images is the truth. However, Sellars (1962) rejects all three of these stances. But the dualism that he dismisses is a dualism of two types of substances, properties or facts. He does not have the dualism that originates in modern subject theory in view. Nonetheless, being a dualism, this stance is based on the commitment to both matter in motion and persons (subjects) as ontological primitives. Eschewing an ontological dualism, the question is whether one can refrain from the commitment to persons as an ontological primitive while retaining the social, normative practices as something that cannot be located in the scientific image.

What Sellars (1962) has in mind as synoptic view is a fourth stance according to which *there are irreducibly normative, social practices situated in the manifest image, but they do not call for an ontological commitment that goes beyond the commitments in the scientific image.* The reason is, again, that norms are not further facts: by recognizing a particle configuration as a person, one adopts a certain attitude to the being in question. Sellars characterizes this stance in the following manner:

> To think of a featherless biped as a person is to think of it as a being with which one is bound up in a network of rights and duties. From this point of view, the irreducibility of the personal is the irreducibility of the 'ought' to the 'is'. But even more basic than this (though, ultimately, as we shall see, the two points coincide), is the fact that to think of a featherless biped as a person is to construe its behaviour in terms of actual or potential membership in an embracing group each member of which thinks of itself as a member of the group. … It follows that to recognize a featherless biped or dolphin or Martian as a person requires that one thinks thoughts of the form, 'We (one) shall do (or abstain from doing) actions of kind A in circumstances C'. To think thoughts of this kind is not to *classify* or *explain*, but to *rehearse an intention*. (Sellars 1962, section VII)

This is a characteristic statement of what has subsequently become known as left-wing Sellarsianism. This stream lays stress on persons being characterized by their participation in social, normative practices, without these practices being something that exists on top of what is recognized in the scientific image; nonetheless, they are not reducible to anything that figures in the scientific image. Prominent philosophers in this stream—with still remarkably different positions—are Richard Rorty (1980) and Robert Brandom (1994).

By contrast, what has subsequently become known as right-wing Sellarsianism[18] is the stream that originates in Sellars's plea for the

---

[18] See e.g. Brandom (2015, pp. 30–32) for explaining the distinction between right-wing and left-wing Sellarsianism.

supremacy of the scientific image when it comes to ontology as expressed, for instance, in his dictum "in the dimension of describing and explaining the world, science is the measure of all things, of what is that it is, and of what is not that it is not" (Sellars 1956, § 41). That stream takes the scientific image to be the complete image of the world in that it considers also what characterizes persons to be reducible to the ingredients of the scientific image. The foremost strategy that is pursued to achieve that reduction in this stream is biological functionalism, as exemplified in the work of Ruth Garrett Millikan (1984). Paul Churchland (1979) and Daniel Dennett (1987) also count as belonging to that stream, although they do not subscribe to biological functionalism.

Talking in terms of left and right wing Sellarsianism is not a political classification. When it comes to politics, left-wing Sellarsianism is not committed to socialism. It can also go with libertarianism, because there are no matters of fact—physical, moral, religious, or otherwise—that could justify restricting the freedom of persons, apart from what is implied by recognizing a being as a member of the community. By the same token, right-wing Sellarsianism is not tied to right-wing political views, even if it takes the form of biological functionalism. The issue is explanation and not to advocate some sort of biological, genetic determinism for society and politics. The classification as "right-wing" expresses only that one lays the stress on scientific realism, and the classification as "left-wing" refers only to the emphasis on social, normative practices.

The question is whether left-wing Sellarsianism is a stable position, if it does not subscribe to a commitment to persons as ontologically primitive. As Sellars says in the quotation above, one may recognize a featherless biped, a dolphin, or a Martian as member of the community. It is just a matter of taking a certain attitude, what Dennett (1987) calls "the intentional stance" or what Sellars characterizes in the quotation above as adopting a certain intention towards the beings in question in contrast to engaging in a classification or explanation of them. Of course, the beings towards which one adopts this stance have to respond in such a way that adopting this stance is not frustrated. Nonetheless, when playing chess with a computer, one can take the attitude of adopting the computer as a member of the community. That attitude is not frustrated as long as one is engaged in the chess game. However, the question whether the computer really has thoughts and follows rules instead of its behaviour merely exhibiting certain regularities makes no sense on this view.

The question therefore is how left-wing Sellarsianism can avoid the consequence of finally eliminating persons. There is no satisfactory answer to this question. Coming back to the methodology of metaphysics set out by means of the quotation from Jackson (1994, p. 25) at the beginning of Sect. 2.1, the standard in metaphysics as in science is this one: if something does not figure explicitly in the ontology originally admitted as primitive, then that something either has to be located in that ontology (that is, one has to show how it figures implicitly in the ontology originally admitted as primitive), or it has to be eliminated, or one has to admit it as a further ontological primitive. Left-wing Sellarsianism starts from the idea that persons can neither be eliminated nor be located in the ontology of the scientific image.

However, stressing the point that admitting a being as a member of a community that is bound by certain rights and obligations "is not to *classify* or *explain*, but to *rehearse an intention*" (Sellars 1962, section VII) cannot hide that in doing so, one subscribes to a substantial ontological commitment, namely the commitment to persons as ontologically primitive. There is no third way between either eliminating something or subscribing to an ontological commitment to it. This then either is a commitment to that something as ontologically primitive or comes with the obligation to show how that something is located in what one admits as ontologically primitive.

Nonetheless, there can be beings that come under different ontological categories. The well-taken point of left-wing Sellarsianism is that persons are not of the same category as matter in motion—or, to put it differently, that the relations individuating matter points are not of the same category as the relations that individuate mind points. Both exist. Both are ontologically primitive. However, recognizing persons over and above matter in motion does not amount to recognizing further substances, properties or facts. One can with good reason maintain that persons exist only in a community of persons such that each member of the community recognizes all the other members as well as herself as persons and that all there is to persons consists in adopting certain attitudes, namely normative ones, towards oneself and the others. Recognizing oneself and others as persons means "that one thinks thoughts of the form, 'We (one) shall do (or abstain from doing) actions of kind A in circumstances C'", as Sellars (1962, section VII) puts it.

In sum, the synoptic view that brings both the scientific and the manifest image together in a "stereoscopic vision, where two differing perspectives on a landscape are fused into one coherent experience" (Sellars 1962, section I) hinges upon spelling out a categorical difference between matter in motion and persons, between facts and norms, without thereby losing out of sight that this vision presupposes an ontological commitment to both. Hence, to the extent that the synoptic view is a stable philosophical position, it is a left-wing Sellarsianism that, however, is motivated by scientific realism and that meets the standards of serious metaphysics. One may therefore say that the view advocated here is a form of what is known as "liberal naturalism" in the current debate[19]— but it is a liberal naturalism that has the ambition to spell out in precise terms how what is not reducible to the ontological primitives of the scientific image is related to these primitives by drawing on Kantian, Sellarsian and Davidsonian resources.

Both the dualism of facts and norms in Kantian or Davidsonian style and left-wing Sellarsianism are akin to existentialism, notably the existentialism set out in Sartre's *Being and Nothingness* (1943).[20] The reason is that on all these views, certain organisms create themselves as persons by adopting normative attitudes towards each other. This is so in existentialism and this is arguably what Kant has in mind when he employs the notion of spontaneity to characterize persons. There may be sufficient physical conditions for the ability to become a person—in other words, sufficient physical conditions for responsiveness to norms; but the exercise of this ability in creating social, normative practices and thereby creating ourselves as persons and the rules and norms that are set up in these practices are not determined by anything physical. They do not supervene on the physical and they are not entailed by the physical description of the world.

---

[19] See De Caro and Voltolini (2010) and De Caro (2015) for that term and a defence of the associated view.

[20] English translation Sartre (1956); see in particular introduction, part IV.1 and conclusion.

The decisive issue is that science gives us laws that, by keeping them fixed, provide a frame for what we can and cannot do physically. But they do not impose certain actions upon us. They do not have the modal force to govern, predetermine or bring about our actions. In general, they cannot fix the rules that persons follow, neither in thoughts, nor in actions. Existentialism is based on the view that there is no such normative system imposed upon us from the world, neither from the physical world nor from a higher-level instance that exists independently of us. We have to create it ourselves and thereby create ourselves as persons. This makes existentialism akin to the dualism of a synoptic view discussed in this section, as well as this dualism akin to existentialism.

The task of philosophy is to develop a reflective equilibrium, that is, gauging which stance achieves a view of the world and our place in it that satisfies best what one expects from such a view. For my part, scientific realism is not negotiable. Science has not only an instrumental value as source of information for our actions. It discovers truth. The activity that Aristotle describes as *theoria* at the beginning of the *Metaphysics* has a value in itself. Kant also expresses this value in employing the expression "the starry heavens above me" in the conclusion of the *Critique of practical reason* (see the quotation above).

On the one hand, this demand on a satisfactory worldview rules the stance out according to which the manifest image is the complete image. On the other hand, scientific realism does not commit us to maintaining that science discovers the whole truth. In particular, I do not see a convincing strategy how to locate meaning and normativity within the scientific image. Considering the attempts to achieve a synoptic view that brings both the scientific and the manifest image together, I do not see how one can get away with recognizing persons without subscribing to an ontological commitment to them. This, then, is a commitment to persons as ontological primitives. But it does not entail that being a person is a further substance, property or fact over and above the physical ones.

The resulting view can be summed up in terms of the following three claims:

(1) *Dualism*: Both matter in motion and persons are ontologically primitive. Both are points that are structurally individuated by the relations

in which they stand. Matter points are individuated by their position in a web of distance relations and the change in these relations. Persons insofar as they are mind points are individuated by their position in a normative web of rights and obligations, commitments, entitlements and precluded entitlements that concerns beliefs as well as actions and that changes through every move that a person makes in her thoughts and actions.

(2) *Categorical difference*: The normative web is categorically distinct from the web of distance relations. It comes into existence only when there are beings that take to themselves and others the attitude of asking the question what they should do and what they should believe. It is accessible only from within participating in the practices that determine this web.

(3) *Coherence*: The scientific image and its method are perfectly coherent and true as far as the account of the matters of fact is concerned. But the very conceptualization, endorsement and justification of this image implies the commitment to persons as ontologically primitive, albeit not on the same footing as matter in motion (i.e. not as a further matter of fact). Therefore, the dualism of persons and matter in motion is the overall most coherent position.

Two ingredients are crucial in order to achieve this overall coherence: The first step is to limit scientific realism to a minimal ontology of the natural world. One does so exclusively for reasons that are situated within scientific realism. Any further ontological commitment undermines scientific realism, because it ends up in the impasse of further problems instead of leading to a gain in explanations. The benefit that is pertinent when it comes to persons then is this one: since the commitment is only to a primitive ontology of matter points individuated by distance relations and their change, all the further dynamical parameters come in through their functional role for this change and thus are located in it. This enables us to maintain that their initial values insofar as they enter into the initial conditions for laws are fixed by the change that actually occurs. Consequently, laws of nature cannot infringe upon the free will of persons and hinder that freedom from manifesting itself in the bodily motions of persons. The second step then is to consider persons not as a further matter of fact. They entirely consist in the normative attitudes that they adopt towards themselves and others. Only in that respect is normativity relevant for this book, namely in order to work out what characterizes us as persons and how this is related to freedom. Normativity then also has of course a much larger significance for society, law, the state, etc., which is beyond the scope of this book.

3.5   A Twofold Conception of Freedom

Let us come back to freedom. The purpose of the considerations about persons in this chapter is to pave the way to a positive conception of freedom, that is, one by means of which persons are distinguished from physical objects with respect to freedom. The argument of Chap. 2 was that there is no threat to freedom from science, even if science gives us fundamental and universal, deterministic laws in physics, regularities in genetics or neuroscience that are stable against the background of normal conditions, etc. The reason is that first comes the motion of the matter in the universe, then come the laws, including the dynamical parameters that enter into the initial conditions of the laws over and above those parameters that define the primitive ontology. Hence, determinism in science does not imply that there is something in the universe that predetermines or even produces the motion of the matter in the universe. Determinism in science is only a thesis about entailment relations between propositions such that the propositions stating the laws and the propositions specifying initial conditions for the laws entail the propositions about the entire past and future evolution of the systems in nature. The motions that actually happen in the universe are sufficient to make these propositions true.

One can therefore say that motion is free in the sense that there are no modal entities in the world that predetermine the motion that actually occurs. The business of science is to discover salient patterns or regularities in the motion of matter such that simple, general and informative laws of nature can be formulated that can then also serve as source of information for our actions. However, free motion in this sense applies to all the objects in the universe, electrons and humans alike. Of course, humans are very complex and highly organized systems in contrast to point particles. Nonetheless, if also human reasoning and deliberating is naturalized in the sense of being located in particle configurations by means of the mentioned functionalist method, then, however coordinated or organized the motion may be, at the end of the day it is motion of particle configurations that just occurs (as, for instance, in the conception of freedom exposed in Ismael 2016).

Hence, if the scientific image is taken to be complete, then the understanding of science in terms of a minimal ontology that leads to Humeanism about laws and/or Super-Humeanism about the dynamical parameters shows that science establishes freedom of motion instead of excluding it; but the freedom is at the end of the day only the contingency of the particle motion that actually occurs. If the scientific image is not taken to be complete, then the understanding of science in this framework shows that there is no obstacle to freedom forthcoming from science; but a positive conception of freedom then has to be formulated in terms of the ontological commitments that lie outside of the scientific image.

The positive conception of freedom that distinguishes human free will from the mere regularities of particle motion consists in the normative attitudes that come with the deliberation about what one should do and think, being subject to a justification in the sense of giving and asking for reasons. This deliberation concerns not only action, but also thought. It creates meaning. It thereby creates ourselves as persons. As Kant puts it, as already quoted,

> If an appearance is given to us, we are still completely free as to how we want to judge things from it.[21]

This is freedom from the realm of matter in motion in the sense that thought and action do not simply follow from sensory impressions and thus particle motion. Taking them to do so would be an instance of what Sellars (1956) denounces as the "myth of the given".

---

[21] *Prolegomena* § 13, note III; quoted from Kant (2002, p. 85); see Sect. 3.2 above.

Persons are free to make up their mind as to what to think and what to do. If thought and action do not follow from particle motion, persons have to set up themselves rules for thought and action. Again, this is the difference between regularities in the sequence of events and the following of rules. By this difference, freedom in thought and action is also distinct from chance events or irregularities: it is the freedom to set up oneself the rules for thought and action. But it is rules. Without rules, there would be no conceptual content, hence no beliefs and no action. Action is based on beliefs and intentions by contrast to a simple reaction to stimuli.

Freedom in thought and action has a positive connotation, as it is the essence of ourselves as persons. However, as again the affinity with existentialism mentioned at the end of the preceding section makes clear, with freedom comes also the responsibility for our thoughts and actions. By setting up rules for thought and action, persons are also responsible for them in the literal sense: they have to answer the question of why these rules (and not others) in terms of giving reasons. This responsibility cannot be shifted to anything else, again on pain of falling into the "myth of the given". It cannot be delegated to sensory experiences or biological needs and inclinations, because they do not impose judgements or actions on us. By contrast, nothing is responsible for the regularities of particle motion and their consequences. They just happen. In short, with normative attitudes and deliberation comes a positive conception of freedom in the guise of a free will that goes with responsibility. This freedom is tied to rules and thereby norms; but these are not automatically moral norms. Moral normativity is just one form of this freedom and responsibility. The issue is in the first place deliberation, rules and thereby normativity in rationality.

This freedom is implemented in the manifest image, which endorses persons as ontologically primitive. By the same token as there is no cogent reason to inflate the ontology of the scientific image with modally loaden entities such as dispositions or powers, which then could lead to a conflict with human freedom, so there is no point in posing non-physical entities beyond the grasp of science that could lead to such a conflict. The freedom of persons in thought and action is the argument for not dismissing the manifest image and taking it to be superseded by science. As elaborated on in Sects. 3.2 and 3.3, any scientific theory, including the scientific image as a whole, is conceived, endorsed and justified in the rule-following practises of persons. The same goes for any theory and any worldview whatsoever.

Also in Christian philosophy, as comes out clearly in Augustinus's *Confessions*, for instance, God is admitted as the origin of everything, but God then creates humans as free beings. Again, persons are ontologically primitive. They are conceived in analogy to God, which assures their freedom. Otherwise, one would again face the *reductio ad absurdum* of the position being formulated through weighing of reasons, which presupposes the very freedom that the position then denies by posing supernatural entities that rule out that freedom.

Descartes and the early modern theories of subjectivity up to and including Kant then endorse not only persons ("cogito, ergo sum") as ontologically primitive, but go as far as seeking to deduce every other knowledge from this primitive. However, as one is not committed in the scientific image to conceive the claims that define the primitive ontology of matter in motion to be the foundation of knowledge, so the endorsement of persons as ontologically primitive does not commit one to regarding the statements about persons as the foundation of knowledge. As mentioned at the end of Sect. 3.2, knowledge claims are justified through coherence. They are justified by integrating them into an overall coherent system of knowledge that in the end encompasses our knowledge about the world as well as about ourselves as those beings who formulate, endorse and justify knowledge claims.

One can thus say that freedom is both the common ground of the scientific and the manifest image as well as what divides these images. It is the common ground in the sense that science gives us contingent laws of motion that leave us freedom for our own bodily motions. It is what divides these images, because the issue is whether this freedom of motion is sufficient to capture the freedom that characterizes thought and action. If and only if one assumes that the latter freedom is distinct from the former freedom, one is committed to a position that endorses persons as ontologically primitive and thereby retains the cornerstone of the manifest image.

Science carves nature at its joints by revealing the salient patterns or regularities in the motion of the matter of the universe. Keeping the laws of physics as well as the regularities of biology, etc. fixed therefore sets a frame within which we can act. One reveals what we can and what we cannot do and what are the likely consequences of chosen courses of action. However, precisely for this reason, namely precisely because of its achievement of objectivity—the point of view from nowhere and nowhen—, science cannot give us norms for our actions. It can only give us facts, namely knowledge of causal or functional relationships, but not what is sometimes described as knowledge of orientation. If scientific theories are given to us, we are still completely free as to how we want to act. We have to deliberate about our actions by weighing reasons. Scientific results cannot *per se* be such reasons. There is no other option than to justify our actions ourselves and bear the responsibility for them. Delegating this responsibility to science is an abuse of science in the same way as delegating this responsibility to religion was an abuse of religion in the prescientific age. The issue can hence not be to let science determine our thoughts and actions. The issue is how to employ the knowledge that stems from the scientific as well as the manifest image to fully develop our capacities as well as our responsibilities as persons.

Summary

The aim of the first chapter is to bring out what the ontological commitments of modern science are and what they are not. Being clear about these commitments is indispensable in order to appreciate why scientific results do not come into conflict with human free will, even if they are formulated in terms of universal and deterministic laws.

(1.1) Atomism is at the roots of the success of modern science, conceiving everything as being composed of microphysical particles and explaining everything in terms of the spatial arrangement of these particles and the change in their arrangement (i.e. their motion).

(1.2) For a science oriented metaphysics, the first and foremost question to be answered is this one: Which ontological commitments are minimally sufficient to understand our scientific knowledge? The claim of this chapter is that there is an answer to this question that covers the whole of modern science from Newtonian mechanics to quantum field theory, as well as from particle physics to biology and neuroscience. The answer is based on atomism. It can be formulated in terms of the following two axioms:

© The Author(s) 2020
M. Esfeld, *Science and Human Freedom*,
https://doi.org/10.1007/978-3-030-37771-7

(1) *There are distance relations that individuate simple objects, namely point particles (matter points).*

(2) *The point particles are permanent, with the distances between them changing.*

These two axioms define what can be called the *primitive ontology* in the sense of what has to be admitted as simply existing in order to understand what science tells us about the world. There is no cogent reason to enrich the ontology beyond these minimal commitments. Doing so only leads to reifying the mathematical structure of scientific theories, ending up in being stuck with pseudo-problems that result from such a reification.

(1.3) Given the commitment to a primitive ontology of matter in motion, the aim of science is to set up a representation of the evolution of the configuration of matter that is as simple and as informative as possible. In order to do so further parameters have to be introduced beyond the primitive ones of distance relations among point particles and their change, since the task is to formulate a representation of that change that is as simple as possible. Hence, simplicity in ontology (minimal sufficient commitments) and simplicity in representation (covering as many motions as possible under as simple a law as possible) pull in opposite directions: parameters are needed that bring out salient patterns or regularities in the motion of matter. These parameters and the laws as well as the geometry formulated by means of them can be considered as the *dynamical structure* of a scientific theory. The dynamical structure changes the more we learn about the motion of matter. The primitive ontology, by contrast, remains constant.

(1.4) A deterministic dynamical structure, if achievable, is the most simple and informative representation, because given the laws and appropriately formulated initial conditions, the whole past and future evolution of the systems under consideration is thereby represented. However, strictly speaking, a deterministic dynamical structure can be conceived only for the universe as a whole. In the case of specific systems—such as e.g. biological determinism—normal environmental conditions have to be presupposed that cannot be specified in a complete and precise manner. For that reason alone, determinism implies nothing about the availability of deterministic predictions. Since initial conditions cannot be known with full precision and since the evolution of systems can be sensitive to slight variations in the initial conditions, deterministic laws have in any case to be linked with a probability measure that enables an answer to the question of which evolution of a system we can expect given the ignorance of its exact initial conditions and those of the environment. The transition from classical mechanics to classical statistical mechanics illustrates how this can be done.

(1.5) Fields as introduced in classical electrodynamics are part of the dynamical structure instead of new elements in the primitive ontology: they are a means to represent retarded particle interactions as local interactions that are mediated by fields.

(1.6) General relativity theory achieves a representation of gravitation as local, retarded interaction. However, there are two possibilities to formulate such a theory: either in terms of a four-dimensional geometry with no privileged foliation into space and time, but a metric of absolute spatio-temporal intervals between point-events; or in terms of successions of instantaneous configurations that are characterized only by relational quantities. This confirms that the four-dimensional geometry belongs to the dynamical structure instead of determining the primitive ontology. The geometry of general relativity theory therefore does not justify drawing the conclusion of the metaphysics of a four-dimensional block universe with no distinction between variation within a configuration and change of the configuration and no temporal becoming. By way of consequence, there is no conflict between science and common sense with respect to time, change and becoming.

(1.7) The dynamical structure of quantum mechanics is radically different from the one of classical mechanics. It raises two problems: the measurement problem, which is the question how a formalism in terms of a wave-function and its evolution that is subject to superpositions and entanglement is related to measurement outcomes and in general matters of fact in the physical world; and the non-locality problem, which results from the mathematically proven fact that any answer to the former question that takes determinate measurement outcomes for granted has to be formulated in terms of a non-local dynamics, that is, a dynamical structure that does not fit into the framework of a local field theory of interactions. The arguably best solution to both these problems is a quantum theory in terms of point particles that move on continuous trajectories, thereby providing for determinate measurement outcomes, but satisfying a non-local law of motion. Such a theory is available for both quantum mechanics and quantum field theory. It confirms the primitive ontology in terms of point particles that are characterized only by the distances in which they stand and the change in these distances.

On this basis, the aim of the second chapter is to formulate an account of scientific laws and explanations that brings out what these explanations achieve and what they do not achieve and thereby to remove the concerns that one may have with respect to human freedom stemming from—universal and deterministic—laws in science.

(2.1) Given a primitive ontology, the problem is how to locate in this ontology all those entities that exist but that do not figure explicitly in the notions that define the primitive ontology. The solution to this problem is functionalism, namely to introduce everything else in terms of a causal or functional role for the evolution of the objects of the primitive ontology. This procedure applies already to the dynamical parameters of physics such as mass, charge, energy, wave-functions, etc. Its most known success is functionalism with respect to the systems and features that the special sciences treat, such as chemistry, biology, neuroscience, etc. One thereby locates everything else in the particle motion in a literal sense: everything else is identical with configurations of matter and their motions. Consequently, a complete description of the universe in terms of the primitive ontology entails all the other true propositions about the natural world. This procedure characterizes what is known as the *scientific image of the world.*

(2.2) Scientific explanations cannot give a reason why the basic particle motion is as it is. Any attempt to do so in philosophy in terms of primitive modal entities such as dispositions, powers or primitive laws results in circular explanations and entails pseudo-problems through its commitment to surplus structure in the ontology. Science seeks to reveal salient patterns or regularities in the basic particle motion (explanation by unification). On the basis of these patterns or regularities, science then provides causal explanations of all the other motions via the mentioned method of functionalization.

(2.3) Laws of nature belong to the dynamical structure of scientific theories. They seek to represent the salient patterns or regularities in the evolution of the objects in their domain. They are true if they identify these patterns. But they do not call for ontological commitments that go beyond the primitive ontology of matter in motion. This stance can be dubbed Super-Humeanism with respect to the dynamical structure of scientific theories.

(2.4) Determinism states that the propositions describing the laws and the initial conditions that enter into the laws entail the propositions that describe the entire past and future evolution of the systems under consideration. Determinism hence does not single out a direction of time. It does not imply that there is something that predetermines, produces or brings about the future evolution of the systems under consideration. Quite to the contrary, one can with good reason maintain the following: first comes the motion of the matter, then come the laws and the dynamical parameters that are needed over and above the primitive ones to determine initial conditions that enter into the laws. Thus, the motion that actually occurs, including the behaviour of persons that is the expression of their free will, fixes the laws and the dynamical parameters that figure in the initial conditions. Super-Humeanism thereby allows us to maintain that if persons decided to do otherwise, the initial values of these parameters would be slightly different. On the one hand, it is thereby excluded that determinism in scientific theories can infringe upon the freedom of persons. On the other hand, laws of nature can still serve as the framework for what persons can and cannot do in their actions.

Against this background, the third chapter seeks to develop a positive conception of free will, over and above removing misconceptions of implications of determinism in science against free will. The third part or chapter considers the interplay between what is known as the *scientific image of the world* and the *manifest image of the world*. The latter is not common sense, but the philosophical stance that puts persons at the centre, regarding them as ontologically primitive.

(3.1) The scientific image faces already problems before it comes to persons, their free will and their rationality. These concern sensory qualities (colours, sounds, smells, tastes) and their perception (qualia, known as the hard problem of consciousness). It is difficult to see how these problems could be solved in a convincing manner by the method of functionalization within the scientific image.

(3.2) The central problems for integrating thought and action into the scientific image concern the transition from syntax to semantics and normativity. Freedom and normativity go together and encompass thought as well as action. Given a sensory input, we are free what to believe in the same way as given sensory inputs as well as biological needs and inclinations, we are free what to do. Thus, the question what one should believe arises in the same way as the question how one should act. Consequently, both what one believes and what one does is subject to a justification in the sense of giving and asking for reasons.

(3.3) If one regards the scientific image as complete also when it comes to persons, their thoughts and actions, the problem is that this image (as well as any scientific theory) is conceived, endorsed and justified in normative attitudes of giving and asking for reasons that presuppose the freedom of persons in formulating, testing and judging theories. In that respect, persons are irreducible and hence ontologically primitive. However, if one endorses persons as the primitive ontology for this reason and takes the manifest image to be complete, the problem is that one is committed to conceiving everything in analogy to persons. One thereby loses the explanatory power of science also when it comes to inanimate matter.

(3.4) The way out of this dilemma is a dualism of both matter in motion and persons being ontological primitives. Both matter and persons are points that are structurally individuated through the relations in which they stand. Matter points are individuated by their position in a web of distance relations. Persons or mind points are individuated by their position in a normative web of rights and obligations, commitments, entitlements and precluded entitlements that concerns beliefs as well as actions. The credibility of such a dualism then hinges upon spelling it out in such a way that being a person is not a further fact (substance, property) over and above the facts (substances, properties) that are described within the scientific image. Being a person is a normative attitude that one adopts to oneself and others and through which one creates oneself as a person by forming beliefs and actions. The normative web of rights and obligations in which beliefs and actions are situated is only accessible from within participating in the normative practices that create it and thereby contributing to further developing it. Facts, by contrast, are accessible from a point of view of nowhere and nowhen. There are sufficient physical conditions for the ability to become a person; but the exercise of this ability in creating ourselves as persons and the rules and norms that are set up in these practices are not determined by anything physical. They do not supervene on the physical and they are not entailed by the physical description of the world.

(3.5) Science and freedom are interwoven in a twofold manner. In the first place, science, even deterministic theories, does not infringe upon freedom: first comes the motion of the matter including the behaviour of humans that is the expression of their free will, then come the laws and the parameters that enter into the initial conditions for the laws. In that sense, any motion is free. Moreover, the very conceptualization, endorsement, testing and justifying of any theory, scientific or otherwise, presupposes the freedom of persons to make up their own minds with respect to what to believe and how to act, given the input that they receive from the world. That freedom cannot be captured within the scientific image. It is not a chance event or an irregularity: it consists in the freedom to set up oneself the rules and thereby the norms for thought and action.

References

Albert, David Z. (2000): *Time and chance*. Cambridge (Massachusetts): Harvard University Press.

Albert, David Z. (2015): *After physics*. Cambridge (Massachusetts): Harvard University Press.

Allori, Valia, Goldstein, Sheldon, Tumulka, Roderich and Zanghì, Nino (2014): "Predictions and primitive ontology in quantum foundations: a study of examples". *British Journal for the Philosophy of Science* 65, pp. 323–352.

Barbour, Julian B. (2003): "Scale-invariant gravity: particle dynamics". *Classical and Quantum Gravity* 20, pp. 1543–1570.

Barbour, Julian B. (2012): "Shape dynamics. An introduction". In: F. Finster, O. Mueller, M. Nardmann, J. Tolksdorf and E. Zeidler (eds.): *Quantum field theory and gravity*. Basel: Birkhaeuser, pp. 257–297.

Barbour, Julian B. and Bertotti, Bruno (1982): "Mach's principle and the structure of dynamical theories". *Proceedings of the Royal Society A* 382, pp. 295–306.

Barbour, Julian B., Koslowski, Tim and Mercati, Flavio (2015): "Entropy and the typicality of universes". *Manuscript*, arXiv:1507.06498 [gr-qc]

Barrett, Jeffrey A. (2014): "Entanglement and disentanglement in relativistic quantum mechanics". *Studies in History and Philosophy of Modern Physics* 47, pp. 168–174.

© The Author(s) 2020
M. Esfeld, *Science and Human Freedom*,
https://doi.org/10.1007/978-3-030-37771-7

Beebee, Helen and Mele, Alfred R. (2002): "Humean compatibilism". *Mind* 111, pp. 201–223.

Bell, John S. (2004): *Speakable and unspeakable in quantum mechanics.* Cambridge: Cambridge University Press. Second edition. First edition 1987.

Belot, Gordon (2001): "The principle of sufficient reason". *Journal of Philosophy* 98, pp. 55–74.

Benovsky, Jiri (2019): *Mind and matter. Panpsychism, dual-aspect monism, and the combination problem.* Cham: Springer.

Bhogal, Harjit (2019): "Nomothetic explanation and Humeanism about laws of nature". Forthcoming in *Oxford Studies in Metaphysics.*

Bhogal, Harjit and Perry, Zee (2017): "What the Humean should say about entanglement". *Noûs* 51, pp. 74–94.

Bilgrami, Akeel (2006): "Some philosophical integrations". In: C. MacDonald and G. MacDonald (eds.): *McDowell and his critics.* Oxford: Blackwell, pp. 50–66.

Bird, Alexander (2007): *Nature's metaphysics. Laws and properties.* Oxford: Oxford University Press.

Bohm, David (1952): "A suggested interpretation of the quantum theory in terms of 'hidden' variables. I and II". *Physical Review* 85, pp. 166–179, 180–193.

Bohm, David (1980): *Wholeness and the implicate order.* London: Routledge.

Bohm, David and Hiley, Basil J. (1993): *The undivided universe. An ontological interpretation of quantum theory.* London: Routledge.

Boltzmann, Ludwig (1896/98): *Vorlesungen über Gastheorie. Teil 1 und 2.* Leipzig: Barth.

Boltzmann, Ludwig (1897): "Zu Hrn. Zermelo's Abhandlung über die mechanische Erklärung irreversibler Vorgänge". *Annalen der Physik* 60, pp. 392–398. English translation "On Zermelo's paper 'On the mechanical explanation of irreversible processes'" in S. G. Brush (1966): *Kinetic theory. Volume 2. Irreversible processes.* Oxford: Pergamon Press, pp. 238–245.

Boltzmann, Ludwig (1964): *Lectures on gas theory.* Translated by Stephen G. Brush. Berkeley: University of California Press.

Brandom, Robert B. (1994): *Making it explicit. Reasoning, representing, and discursive commitment.* Cambridge (Massachusetts): Harvard University Press.

Brandom, Robert B. (2015): *From empiricism to expressivism. Brandom reads Sellars.* Cambridge (Massachusetts): Harvard University Press.

Brennan, Jason (2007): "Free will in the block universe". *Philosophia* 35, pp. 207–217.

Breuer, Thomas (1995): "The impossibility of exact state self-measurements". *Philosophy of Science* 62, pp. 197–214.

Bricmont, Jean (2016): *Making sense of quantum mechanics.* Cham: Springer.

Brown, Harvey R., Dewdney, Chris and Horton, G. (1995): "Bohm particles and their detection in the light of neutron interferometry". *Foundations of Physics* 25, pp. 329–347.

Brown, Harvey R., Elby, Andrew and Weingard, Robert (1996): "Cause and effect in the pilot-wave interpretation of quantum mechanics". In: J. T. Cushing, A. Fine and S. Goldstein (eds.): *Bohmian mechanics and quantum theory: an appraisal.* Dordrecht: Springer, pp. 309–319.

Brüntrup, Godehard and Jaskolla, Ludwig (eds.) (2017): *Panpsychism. Contemporary perspectives.* Oxford: Oxford University Press.

Callender, Craig (2004): "Measures, explanations and the past: should 'special' initial conditions be explained?". *British Journal for the Philosophy of Science* 55, pp. 195–217.

Callender, Craig (2015): "One world, one beable". *Synthese* 192, pp. 3153–3177.

Carnap, Rudolf (1928): *Scheinprobleme in der Philosophie. Das Fremdpsychische und der Realismusstreit.* Berlin-Schlachtensee: Weltkreis Verlag.

Carroll, Sean (2010): *From eternity to here. The quest for the ultimate theory of time.* New York: Penguin.

Chalmers, David J. (1996): *The conscious mind. In search of a fundamental theory.* Oxford: Oxford University Press.

Chen, Eddy Keming (2019): "Quantum mechanics in a time-asymmetric universe: on the nature of the initial quantum state". Forthcoming in the *British Journal for the Philosophy of Science.* Preprint http://arxiv.org/abs/1712.01666 [quant-ph]

Churchland, Paul M. (1979): *Matter and consciousness.* Cambridge (Massachusetts): MIT Press.

Colin, Samuel and Struyve, Ward (2007): "A Dirac sea pilot-wave model for quantum field theory". *Journal of Physics A* 40, pp. 7309–7341.

Correia, Fabrice and Schnieder, Benjamin (eds.) (2012): *Metaphysical grounding.* Cambridge: Cambridge University Press.

Cowan, Charles Wesley and Tumulka, Roderich (2016): "Epistemology of wave function collapse in quantum physics". *British Journal for the Philosophy of Science* 67, pp. 405–434.

Darby, George (2018): "A minimalist Humeanism?". *Metasience* 27, pp. 433–437.

Davidson, Donald (1970): "Mental events". In: L. Foster and J. W. Swanson (eds.): *Experience and theory*. Amherst: University of Massachusetts Press, pp. 79–101.

Davidson, Donald (1984): *Inquiries into truth and interpretation*. Oxford: Oxford University Press.

Davidson, Donald (1991): "Three varieties of knowledge". In: A. Philipps Griffiths (ed.): *A. J. Ayer memorial essays. Royal Institute of Philosophy Supplement* 30. Cambridge: Cambridge University Press, pp. 153–166.

de Broglie, Louis (1928): "La nouvelle dynamique des quanta". In: *Electrons et photons. Rapports et discussions du cinquième Conseil de physique tenu à Bruxelles du 24 au 29 octobre 1927 sous les auspices de l'Institut international de physique Solvay*. Paris: Gauthier-Villars, pp. 105–132. English translation in G. Bacciagaluppi and A. Valentini (2009): *Quantum theory at the crossroads. Reconsidering the 1927 Solvay conference*. Cambridge: Cambridge University Press, pp. 341–371.

De Caro, Mario and Voltolini, Alberto (2010): "Is liberal naturalism possible?" In: M. De Caro and D. MacArthur (eds.): *Naturalism and normativity*. New York: Columbia University Press, pp. 69–86.

De Caro, Mario (2015): "Realism, common sense, and science". *The Monist* 98, pp. 1–18.

Deckert, Dirk-André and Hartenstein, Vera (2016): "On the initial value formulation of classical electrodynamics". *Journal of Physics A* 49, pp. 445202–445221.

Dennett, Daniel C. (1987): *The intentional stance*. Cambridge (Massachusetts): MIT Press.

Dennett, Daniel C. (1991): *Consciousness explained*. London: Penguin.

Dowker, Fay and Herbauts, Isabelle (2005): "The status of the wave function in dynamical collapse models". *Foundations of Physics Letters* 18, pp. 499–518.

Dürr, Detlef, Goldstein, Sheldon and Zanghì, Nino (2013): *Quantum physics without quantum philosophy*. Berlin: Springer.

Dürr, Detlef, Goldstein, Sheldon and Zanghì, Nino (2018): "Quantum motion on shape space and the gauge dependent emergence of dynamics and probability in absolute space and time". http://arxiv.org/abs/1808.06844 [quant-ph]

Dürr, Detlef and Teufel, Stefan (2009): *Bohmian mechanics. The physics and mathematics of quantum theory*. Berlin: Springer.

Einstein, Albert (1948): "Quanten–Mechanik und Wirklichkeit". *Dialectica* 2, pp. 320–324.

Esfeld, Michael (2001): *Holism in philosophy of mind and philosophy of physics*. Dordrecht: Kluwer.

Esfeld, Michael (2004): "Quantum entanglement and a metaphysics of relations". Studies in History and Philosophy of Modern Physics 35, pp. 601–617.

Esfeld, Michael (2014a): "The primitive ontology of quantum physics: guidelines for an assessment of the proposals". *Studies in History and Philosophy of Modern Physics* 47, pp. 99–106.

Esfeld, Michael (2014b): "Quantum Humeanism". *Philosophical Quarterly* 64, pp. 453–470.

Esfeld, Michael (2015): "Bell's theorem and the issue of determinism and indeterminism". *Foundations of Physics* 45, pp. 471–482.

Esfeld, Michael and Deckert, Dirk-André (2017): *A minimalist ontology of the natural world*. New York: Routledge.

Esfeld, Michael and Gisin, Nicolas (2014): "The GRW flash theory: a relativistic quantum ontology of matter in space-time?". *Philosophy of Science* 81, pp. 248–264.

Esfeld, Michael and Lam, Vincent (2008): "Moderate structural realism about space-time". *Synthese* 160, pp. 27–46.

Esfeld, Michael, Lazarovici, Dustin, Lam, Vincent and Hubert, Mario (2017): "The physics and metaphysics of primitive stuff". *British Journal for the Philosophy of Science* 68, pp. 133–161.

Esfeld, Michael and Sachse, Christian (2011): *Conservative reductionism*. New York: Routledge.

Everett, Hugh (1957): "'Relative state' formulation of quantum mechanics". *Reviews of Modern Physics* 29, pp. 454–462.

Feynman, Richard P. (1966): "The development of the space-time view of quantum electrodynamics. Nobel Lecture, December 11, 1965". *Science* 153, pp. 699–708.

Feynman, Richard P., Leighton, Robert B. and Sands, Matthew (1963): *The Feynman lectures on physics. Volume 1*. Reading (Massachusetts): Addison-Wesley.

Field, Hartry H. (1980): *Science without numbers. A defence of nominalism*. Oxford: Blackwell.

Fodor, Jerry A. (1987): *Psychosemantics. The problem of meaning in the philosophy of mind*. Cambridge (Massachusetts): MIT Press.

Forrest, Peter (1985): "Backward causation in defence of free will". *Mind* 94, pp. 210–217.

Frankfurt, Harry G. (1971): "Freedom of the will and the concept of a person". *Journal of Philosophy* 68, pp. 5–20.

French, Steven (2014): *The structure of the world. Metaphysics and representation.* Oxford: Oxford University Press.

French, Steven and Ladyman, James (2003): "Remodelling structural realism: quantum physics and the metaphysics of structure". *Synthese* 136, pp. 31–56.

Friedman, Michael (1974): "Explanation and scientific understanding". *Journal of Philosophy* 71, pp. 5–19.

Geach, Peter (1965): "Some problems about time". *Proceedings of the British Academy* 51, pp. 321–336.

Ghirardi, Gian Carlo, Grassi, Renata and Benatti, Fabio (1995): "Describing the macroscopic world: closing the circle within the dynamical reduction program". *Foundations of Physics* 25, pp. 5–38.

Ghirardi, Gian Carlo, Rimini, Alberto and Weber, Tullio (1986): "Unified dynamics for microscopic and macroscopic systems". *Physical Review D* 34, pp. 470–491.

Gillet, Carl (2016): *Reduction and emergence in science and philosophy.* Oxford: Oxford University Press.

Gisin, Nicolas (1984): "Quantum measurements and stochastic processes". *Physical Review Letters* 52, pp. 1657–1660.

Goldstein, Sheldon (2017): "Bohmian mechanics". In: E. N. Zalta (ed.): *The Stanford Encyclopedia of Philosophy* (Summer 2017 edition). https://plato.stanford.edu/archives/sum2017/entries/qm-bohm/

Goldstein, Sheldon, Norsen, Travis, Tausk, Daniel Victor and Zanghì, Nino (2011): "Bell's theorem". http://www.scholarpedia.org/article/Bell's_theorem

Goldstein, Sheldon, Taylor, James, Tumulka, Roderich and Zanghì, Nino (2005a): "Are all particles real?". *Studies in History and Philosophy of Modern Physics* 36, pp. 103–112.

Goldstein, Sheldon, Taylor, James, Tumulka, Roderich and Zanghì, Nino (2005b): "Are all particles identical?". *Journal of Physics A* 38, pp. 1567–1576.

Gomes, Henrique, Gryb, Sean and Koslowski, Tim (2011): "Einstein gravity as a 3d conformally invariant theory". *Classical and Quantum Gravity* 28, p. 045005.

Gomes, Henrique and Koslowski, Tim (2013): "Frequently asked questions about shape dynamics". *Foundations of Physics* 43, pp. 1428–1458.

Graham, Daniel W. (2010): *The texts of early Greek philosophy. The complete fragments and selected testimonies of the major Presocratics.* Edited and translated by Daniel W. Graham. Cambridge: Cambridge University Press.

Gryb, Sean and Thébault, Karim P. Y. (2016): "Time remains". *British Journal for the Philosophy of Science* 67, pp. 663–705.

Hacking, Ian (1975): "The identity of indiscernibles". *Journal of Philosophy* 72, pp. 249–256.

Hall, Ned (2009): "Humean reductionism about laws of nature". Unpublished manuscript, http://philpapers.org/rec/HALHRA

Hartenstein, Vera and Hubert, Mario (2019): "When fields are not degrees of freedom". Forthcoming in the *British Journal for the Philosophy of Science*. Preprint http://philsci-archive.pitt.edu/14911/

Hoefer, Carl (2002): "Freedom from the inside out". *Royal Institute of Philosophy Supplement* 50, pp. 201–222.

Howard, Don (1985): "Einstein on locality and separability". *Studies in History and Philosophy of Science* 16, pp. 171–201.

Hoyningen-Huene, Paul (2013): *Systematicity. The nature of science*. Oxford: Oxford University Press.

Hüttemann, Andreas and Loew, Christian (2019): "Freier Wille und Naturgesetze – Überlegungen zum Konsequenzargument". In: K. von Stoch, S. Wendel, M. Breul and A. Langenfeld (eds.): *Streit um die Freiheit – philosophische und theologische Perspektiven*. Mentis: Paderborn, pp. 77–93.

Ismael, Jenann (2016): *How physics makes us free*. Oxford: Oxford University Press.

Jackson, Frank (1994): "Armchair metaphysics". In: J. O'Leary-Hawthorne and M. Michael (eds.): *Philosophy in mind*. Dordrecht: Kluwer, pp. 23–42.

Jackson, Frank (1998): *From metaphysics to ethics. A defence of conceptual analysis*. Oxford: Oxford University Press.

Kant, Immanuel (1983): *Perpetual peace and other essays on politics, history, and morals*. Translated by Ted Humphrey. Indianapolis: Hackett.

Kant, Immanuel (1996): *The Cambridge edition of the works of Immanuel Kant. Volume 4. Practical philosophy*. Edited by Mary J. Gregor. Cambridge: Cambridge University Press.

Kant, Immanuel (2002): *The Cambridge edition of the works of Immanuel Kant. Volume 3. Theoretical philosophy after 1781*. Edited by Henry Allison and Peter Heath. Cambridge: Cambridge University Press.

Kim, Jaegwon (1998): *Mind in a physical world. An essay on the mind-body problem and mental causation*. Cambridge (Massachusetts): MIT Press.

Kim, Jaegwon (2005): *Physicalism, or something near enough*. Princeton: Princeton University Press.

Kitcher, Philip (1989): "Explanatory unification and the causal structure of the world". In: P. Kitcher and W. C. Salmon (eds.): *Minnesota Studies in the philosophy of science. Volume XIII: Scientific explanation*. Minneapolis: University of Minnesota Press, pp. 410–505.

Koslowski, Tim (2017): "Quantum inflation of classical shapes". *Foundations of Physics* 47, pp. 625–639.

Kripke, Saul A. (1982): *Wittgenstein on rules and private language.* Oxford: Blackwell.

Ladyman, James (1998): "What is structural realism?". *Studies in History and Philosophy of Modern Science* 29, pp. 409–424.

LaMettrie, Julien Offray de (1747): *L'homme machine.* Leyden.

Lance, Mark and O'Leary-Hawthrone, John (1997): *The grammar of meaning: normativity and semantic discourse.* Cambridge: Cambridge University Press.

Lange, Marc (2002): *An introduction to the philosophy of physics: locality, fields, energy and mass.* Oxford: Blackwell.

Laplace, Pierre Simon (1951): *A philosophical essay on probabilities.* Translated by F. W. Truscott and F. L. Emory. New York: Dover.

Lazarovici, Dustin (2018a): "Against fields". *European Journal for the Philosophy of Science* 8, pp. 145–170.

Lazarovici, Dustin (2018b): "Super-Humeanism: a starving ontology". *Studies in History and Philosophy of Modern Physics* 64, pp. 79–86.

Lazarovici, Dustin, Oldofredi, Andrea and Esfeld, Michael (2018): "Observables and unobservables in quantum mechanics: How the no-hidden-variables theorems support the Bohmian particle ontology". *Entropy* 20, pp. 381–397.

Lazarovici, Dustin and Reichert, Paula (2015): "Typicality, irreversibility and the status of macroscopic laws". *Erkenntnis* 80, pp. 689–716.

Lazarovici, Dustin and Reichert, Paula (2019): "Arrow(s) of time without a past hypothesis". http://arxiv.org/abs/1809.04646 [physics.hist-ph]

Leibniz, Gottfried Wilhelm (1890): *Die philosophischen Schriften von G. W. Leibniz. Band 7.* Edited by C. I. Gerhardt. Berlin: Weidmannsche Verlagsbuchhandlung.

Leibniz, Gottfried Wilhelm (2000): *G. W. Leibniz and S. Clarke: Correspondence.* Edited by Roger Ariew. Indianapolis: Hackett.

Levine, Joseph (1983): "Materialism and qualia: The explanatory gap". *Pacific Philosophical Quarterly* 64, pp. 354–361.

Lewis, David (1966): "An argument for the identity theory". *Journal of Philosophy* 63, pp. 17–25.

Lewis, David (1970): "How to define theoretical terms". *Journal of Philosophy* 67, pp. 427–446.

Lewis, David (1972): "Psychophysical and theoretical identifications". *Australasian Journal of Philosophy* 50, pp. 249–258.

Lewis, David (1981): "Are we free to break the laws?". *Theoria* 47, pp. 113–121.

Lewis, David (1986a): *On the plurality of worlds.* Oxford: Blackwell.

Lewis, David (1986b): *Philosophical papers. Volume 2.* Oxford: Oxford University Press.

Lewis, David (1994): "Humean supervenience debugged". *Mind* 103, pp. 473–490.

Lewis, David (2009): "Ramseyan humility". In: D. Braddon-Mitchell and R. Nola (eds.): *Conceptual analysis and philosophical naturalism.* Cambridge (Massachusetts): MIT Press, pp. 203–222.

Libet, Benjamin (2004): *Mind time. The temporal factor in consciousness.* Cambridge (Massachusetts): Harvard University Press.

Loewer, Barry (1996): "Freedom from physics: quantum mechanics and free will". *Philosophical Topics* 24, pp. 91–112.

Loewer, Barry (2007): "Laws and natural properties". *Philosophical Topics* 35, pp. 313–328.

Loewer, Barry (2012): "Two accounts of law and time". *Philosophical Studies* 160, pp. 115–137.

Mach, Ernst (1919): *The science of mechanics: a critical and historical account of its development.* Fourth edition. Translation by Thomas J. McCormack. Chicago: Open Court.

Marmodoro, Anna (2018): "Atomism, holism and structuralism: costs and benefits of a minimalist ontology of the world". *Metasience* 27, pp. 421–425.

Marmodoro, Anna (ed.) (2010): *The metaphysics of powers: their grounding and their manifestations.* London: Routledge.

Matarese, Vera (2019): "A challenge for Super-Humeanism: the problem of immanent comparisons". Forthcoming in *Synthese*, https://doi.org/10.1007/s11229-018-01914-y

Maudlin, Tim (1995): "Three measurement problems". *Topoi* 14, pp. 7–15.

Maudlin, Tim (2002): "Remarks on the passing of time". *Proceedings of the Aristotelian Society* 102, pp. 237–252.

Maudlin, Tim (2007): *The metaphysics within physics.* Oxford: Oxford University Press.

Maudlin, Tim (2010): "Can the world be only wavefunction?". In: S. Saunders, J. Barrett, A. Kent and D. Wallace (eds.): *Many worlds? Everett, quantum theory, and reality.* Oxford: Oxford University Press, pp. 121–143.

Maudlin, Tim (2011): *Quantum non-locality and relativity.* Chichester: Wiley-Blackwell. Third edition. First edition 1994.

Maudlin, Tim (2012): *Philosophy of physics. Space and time.* Princeton: Princeton University Press.

Maudlin, Tim (2019): *Philosophy of physics. Quantum theory.* Princeton: Princeton University Press.

McDowell, John (1995): "Two sorts of naturalism". In: R. Hursthouse, G. Lawrence and W. Quinn (eds.): *Virtues and reasons: Philippa Foot and moral theory.* Oxford: Oxford University Press, pp. 149–179.

Mele, Alfred R. (2014): *Free: why science hasn't disproved free will.* Oxford: Oxford University Press.

Mercati, Flavio (2018): *Shape dynamics: relativity and relationalism.* Oxford: Oxford University Press.

Miller, Elizabeth (2014): "Quantum entanglement, Bohmian mechanics, and Humean supervenience". *Australasian Journal of Philosophy* 92, pp. 567–583.

Millikan, Ruth Garrett (1984): *Language, thought, and other biological categories.* Cambridge (Massachusetts): MIT Press.

Mumford, Stephen and Anjum, Rani Lill (2011): *Getting causes from powers.* Oxford: Oxford University Press.

Newton, Isaac (1934): *Sir Isaac Newton's mathematical principles of natural philosophy and his system of the world.* Translated by Andrew Motte in 1729, revised by Florian Cajori. Volume 1: The motion of bodies. Berkeley: University of California Press.

Newton, Isaac (1952): *Opticks or a treatise of the reflections, refractions, inflections and colours of light.* Edited by I. B. Cohen. New York: Dover.

Newton, Isaac (1961): *Correspondence. Volume II.* Edited by H. W. Turnbull. Cambridge: Cambridge University Press.

Papineau, David (2002): *Thinking about consciousness.* Oxford: Oxford University Press.

Popper, Karl (1945): *The open society and its enemies.* London: Routledge.

Popper, Karl (1950a): "Indeterminism in quantum physics and in classical physics. Part I". *British Journal for the Philosophy of Science* 1, pp. 117–133.

Popper, Karl (1950b): "Indeterminism in quantum physics and in classical physics. Part II". *British Journal for the Philosophy of Science* 1, pp. 173–195.

Popper, Karl (1980): "Three worlds". In: S. M. McMurrin (ed.): *The Tanner Lectures on human values.* Cambridge: Cambridge University Press, pp. 141–167.

Price, Huw (2004): "Naturalism without representationalism". In: M. de Caro and D. Macarthur (eds.): *Naturalism in question.* Cambridge (Massachusetts): Harvard University Press, pp. 71–88.

Putnam, Hilary (1975): "The meaning of 'meaning'". In: H. Putnam: *Mind, Language and Reality. Philosophical Papers Volume 2.* Cambridge: Cambridge University Press, pp. 215–271.

Pylkkänen, Paavo, Hiley, Basil J. and Pättiniemi, Ilkka (2015): "Bohm's approach and individuality". In: A. Guay and T. Pradeu (eds.): *Individuals across the sciences*. Oxford: Oxford University Press, pp. 226–246.

Rorty, Richard (1980): *Philosophy and the mirror of nature*. Oxford: Blackwell.

Rovelli, Carlo (1997): "Halfway through the woods: contemporary research on space and time". In: J. Earman and J. Norton (eds.): *The cosmos of science*. Pittsburgh: University of Pittsburgh Press, pp. 180–223.

Russell, Bertrand (1912): "On the notion of cause". *Proceedings of the Aristotelian Society* 13, pp. 1–26.

Russell, Bertrand (1914): "The problem of infinity considered historically". In: B. Russell: *Our knowledge of the external world*. Chicago: Open Court. Ch. 6.

Sartre, Jean-Paul (1943): *L'être et le néant*. Paris: Gallimard.

Sartre, Jean-Paul (1956): *Being and nothingness*. Translation by Hazel Barnes. London: Routledge.

Scanlon, T. M. (2014): *Being realistic about reasons*. Oxford: Oxford University Press.

Searle, John R. (1980): "Minds, brains, and programs". *Behavioral and Brain Sciences* 3, pp. 417–424, 450–457.

Searle, John R. (1997): *The mystery of consciousness and exchanges with Daniel C. Dennett and David J. Chalmers*. New York: The New York Review of Books.

Seibt, Johanna (1990): *Properties as processes. A synoptic study in W. Sellars' nominalism*. Ridgeview: Reseda.

Sellars, Wilfrid (1956): "Empiricism and the philosophy of mind". In: H. Feigl and M. Scriven (eds.): *The foundations of science and the concepts of psychology and psychoanalysis*. Minneapolis: University of Minnesota Press, pp. 253–329.

Sellars, Wilfrid (1962): "Philosophy and the scientific image of man". In: R. Colodny (ed.): *Frontiers of science and philosophy*. Pittsburgh: University of Pittsburgh Press, pp. 35–78.

Sellars, Wilfrid (1981): "Foundations for a metaphysics of pure processes: I. The lever of Archimedes. II. Naturalism and process. III. Is consciousness physical?". *The Monist* 64, pp. 3–90.

Simpson, William M. R. (2019): "What's the matter with Super-Humeanism?". Forthcoming in *British Journal for the Philosophy of Science*, https://doi.org/10.1093/bjps/axz028

Smith, Michael (1994): *The moral problem*. Oxford: Blackwell.

Strawson, Galen (1989): *The secret connexion. Causation, realism, and David Hume*. Oxford: Oxford University Press.

Strawson, Galen (2017): "Mind and being. The primacy of panpsychism". In: G. Brüntrup and L. Jaskolla (eds.): *Panpsychism. Contemporary perspectives*. Oxford: Oxford University Press, pp. 75–112.

Swartz, Norman (2003): *The concept of physical law*. https://www.sfu.ca/~swartz/physical-law/index.htm. First edition Cambridge: Cambridge University Press 1985.

Tomassello, Michael (2014): *A natural history of human thinking*. Cambridge (Massachusetts): Harvard University Press.

Tumulka, Roderich (2006): "A relativistic version of the Ghirardi-Rimini-Weber model". *Journal of Statistical Physics* 125, pp. 825–844.

van Brakel, Jaap (1996): "Interdiscourse or supervenience relations: the primacy of the manifest image". *Synthese* 106, pp. 253–297.

van Inwagen, Peter (1975): "The incompatibility of free will and determinism". *Philosophical Studies* 27, pp. 185–199.

van Inwagen, Peter (1983): *An essay on free will*. Oxford: Oxford University Press.

Vassallo, Antonio (2015): "Can Bohmian mechanics be made background independent?". *Studies in History and Philosophy of Science* 52, pp. 242–250.

Vassallo, Antonio and Ip, Pui Him (2016): "On the conceptual issues surrounding the notion of relational Bohmian dynamics". *Foundations of Physics* 46, pp. 943–972.

von Wachter, Daniel (2015): "Miracles are not violations of the laws of nature because the laws do not entail regularities". *European Journal for Philosophy of Religion* 7, pp. 37–60.

Wallace, David (2008): "Philosophy of quantum mechanics". In: D. Rickles (ed.): *The Ashgate companion to contemporary philosophy of physics*. Aldershot: Ashgate, pp. 16–98.

Wallace, David (2012): *The emergent multiverse. Quantum theory according to the Everett interpretation*. Oxford: Oxford University Press.

Watson, James D. and Crick, Francis H. C. (1953): "A structure for deoxyribose nucleic acid". *Nature* 171, pp. 737–738.

Weyl, Hermann (1949): *Philosophy of mathematics and natural science*. Princeton: Princeton University Press.

Wheeler, John A. and Feynman, Richard P. (1945): "Interaction with the absorber as the mechanism of radiation". *Reviews of Modern Physics* 17, pp. 157–181.

Wilson, Alastair (2018): "Super-Humeanism: insufficiently naturalistic and insufficiently explanatory". *Metascience* 27, pp. 427–431.

Wittgenstein, Ludwig (1953): *Philosophical Investigations*. Translated by G. E. M. Anscombe. Oxford: Blackwell.

Index[1]

A

Absolute space, 8–10, 12–14, 18, 31, 34, 36, 37, 43, 60, 80, 109
Action at a distance, 16, 29, 38, 57
Albert, David Z., 26–28, 48
Allori, Valia, 51
Analogy, 133, 136–140, 160
Anaxagoras, 5
Anaximander, 5
Anaximenes, 5
Anjum, Rani Lill, 78n6, 94
Anomalous monism, 144, 146, 149
Aristotle, v, 4, 126, 133, 155
Atomism, 1–6, 9, 12, 17, 24, 25, 32, 53, 61, 64, 65
Augustinus, 160

B

Barbour, Julian B., 11, 11n9, 14, 44, 83–85
Barrett, Jeffrey A., 59n38, 60n40
Beebee, Helen, 98n18
Bell, John S., 3, 3n1, 49, 51, 53, 55n33, 58
Bell's theorem, 55, 56, 58, 59
Belot, Gordon, 8n3
Benovsky, Jiri, 138n11
Bertotti, Bruno, 14n10
Best system, 87, 89, 129
Bhogal, Harjit, 77n5, 82n7
Bilgrami, Akeel, 143n13
Bird, Alexander, 78n6, 94
Block universe, 41–43, 59, 103, 108, 109

[1] Note: Page numbers followed by 'n' refer to notes.

© The Author(s) 2020
M. Esfeld, *Science and Human Freedom*,
https://doi.org/10.1007/978-3-030-37771-7

Bohm, David, 49, 51–53, 71, 101,
    115, 115n4
Bohmian mechanics, 49, 50, 53, 54,
    58, 59, 71, 72
Boltzmann, Ludwig, 25, 26
Brandom, Robert B., 118n6,
    125–128, 136, 148, 151,
    151n18
Brennan, Jason, 103n20
Breuer, Thomas, 20n13
Bricmont, Jean, 49n24
Brown, Harvey R., 71n3, 88n12
Brüntrup, Godehard, 138n11

C

Callender, Craig, 82n7, 84n8, 86
Carnap, Rudolf, x, 61, 109
Carroll, Sean, 85n9
Causation, 18, 36, 70, 107, 150
Chalmers, David J., 114
Chen, Eddy Keming, 57n35, 84n8
Churchland, Paul M., 152
Classical mechanics, 2, 14, 22, 24,
    25, 29–36, 45, 46, 54, 61, 83,
    92, 93, 107
Coherence, 85, 156, 160
Colin, Samuel, 60n41
Collapse of the wave-function, 49,
    50, 93
Compatibilism, 96–98, 106
Configuration space, 18, 54, 55, 82,
    101
Consequence argument, 95–98, 101,
    105, 106
Correia, Fabrice, 145n15
Cowan, Charles Wesley, 48, 51n28
Crick, Francis H. C., 66

D

Darby, George, 87n10
Davidson, Donald, 126, 128,
    144–146, 148, 149, 154
de Broglie, Louis, 49, 53
De Caro, Mario, 154n19
Deckert, Dirk-André, 9, 35, 38n20,
    60n41, 82, 87, 92n15
Dennett, Daniel C., 121n7, 152
Descartes, René, 4, 8, 9, 9n5, 111,
    111n1, 116, 117, 125, 140,
    141, 160
Determinism, 50, 56, 92–109, 128,
    129, 149, 152, 157
Dispositions, 78–80, 91, 94, 95,
    106, 159
Double slits experiment, 46, 47, 51,
    54, 55
Dowker, Fay, 82n7
Dualism, 140, 141, 144, 149, 150,
    154–156
Dürr, Detlef, 49n24, 50n26, 58n36,
    60n39, 147n17
Dynamical structure, 9–21, 29,
    33–35, 37, 43, 44, 60, 61, 66,
    81, 90, 92, 93, 100, 112, 118

E

Einstein, Albert, vii, 32, 37–39, 44,
    58
Electrodynamics, 30, 32, 36, 37,
    38n20, 40, 43, 55, 57–59, 61,
    68, 83, 107
Emergence, v, 74, 107, 138, 143
Enlightenment, v, vi, ix, 86
Entanglement, 58, 71
Entropy, 25–28, 41, 42, 73, 83–85

Esfeld, Michael, 9, 70, 73, 82n7, 86, 87
Essence, ix, 4, 7, 88, 133, 137, 159
Everett, Hugh, 48
Evolutionary biology, vi, vii, 93
Existentialism, 154, 155, 159
Explanatory gap, 118, 119, 131

F

Feynman, Richard P., 1–4, 33, 34, 60n42, 67
Field, vii, 3, 19, 29–44, 51–55, 57, 60, 86, 113
Field, Hartry H., 31
Flashes, 51, 52
Fodor, Jerry A., 121
Forrest, Peter, 102
Frankfurt, Harry G., 96
French, Steven, 8n4, 70
Friedman, Michael, 77n5
Functional definition, viii, 19, 68–70, 72, 73, 75, 76, 78, 81, 100, 112, 114, 117–119, 123, 130, 134
Functionalism, 63–75, 120–123, 135, 136, 152

G

Geach, Peter, 41
General relativity, vii, 40, 42, 44, 57, 59
Ghirardi, Gian Carlo, 50, 51
Gillet, Carl, 107
Gisin, Nicolas, 50n27, 59n38
God, 1, 125, 140, 143, 147, 160
Goldstein, Sheldon, 49n24, 55n33, 72n4

Gomes, Henrique, 44n21
Graham, Daniel W., 1
Gravitational waves, 3, 33
Grounding, 145, 146
GRW theory, 50, 93
Gryb, Sean, 44n21

H

Hacking, Ian, 8n3
Hall, Ned, 86
Hamilton, William Rowan, 23, 24
Hartenstein, Vera, 35, 35n18, 38n20, 82
Heidegger, Martin, 125
Heisenberg uncertainty relations, 45, 46, 147, 147n17
Herbauts, Isabelle, 82n7
Hidden variables, 49
Hiley, Basil J., 115n4
Hoefer, Carl, 103, 103n20
Holism, 127, 128, 144
Howard, Don, 58n37
Hoyningen-Huene, Paul, 87n11, 107n21
Hubert, Mario, 35n18, 38n20, 82
Hume, David, 86, 92n15
Humeanism, 86–91, 98–100, 103–106, 109, 158
Humean metaphysics, 86, 90, 92, 106
Hüttemann, Andreas, 100

I

Identity of indiscernibles, 8
Impenetrability, 8
Individuation, 8, 17
Inertial motion, 13, 14, 18

Initial conditions, 15, 16, 18, 20–29,
    34, 39, 45, 51, 52, 54, 56, 59,
    61, 73, 84, 85, 88, 90, 92–94,
    96–98, 100–106, 108, 109,
    150, 156, 157
Ip, Pui Him, 60n39
Ismael, Jenann, 98, 103, 103n20,
    158

J

Jackson, Frank, 63, 64, 64n1, 72, 74,
    88, 112n2, 123, 153
Jaskolla, Ludwig, 138n11
Justification, 121–123, 125, 126,
    129, 130, 135, 146, 148, 156

K

Kant, Immanuel, v, vn1, ix, 124,
    124n8, 128, 144, 144n14,
    154, 155, 158, 158n21, 160
Kim, Jaegwon, 66, 120, 145, 149
Kitcher, Philip, 77n5
Koslowski, Tim, 44n21, 60n39
Kripke, Saul A., 126, 128

L

La Mettrie, Julien Offray de, v, vi
Ladyman, James, 8n4, 70
Lam, Vincent, 8n4, 70
Lance, Mark, 147n16
Lange, Marc, 17n11
Laplace, Pierre Simon, 19, 20, 20n12
Laws of nature, 62, 69, 81, 86–92,
    94–102, 104–108, 129, 156,
    157

Lazarovici, Dustin, 11n8, 23n14,
    31n15, 33n17, 35n18, 53n32,
    85n9, 87n10
Leibniz, Gottfried Wilhelm, 8, 10,
    10n6, 10n7, 12, 14, 17, 28,
    36, 39, 42, 80, 140, 141
Levine, Joseph, 118
Lewis, David, 7, 65, 72, 86, 88–90,
    88n14, 98, 106
Liberal naturalism, 154
Libertarianism, 152
Libet, Benjamin, 132
Light cone, 37, 40, 43, 55, 56, 58
Local action, 58
Locality, 56, 57
Locate/location, 52, 57, 63–75, 81,
    82, 91, 104, 105, 112–114,
    116–118, 121–123, 130, 135,
    136, 145, 155
Loew, Christian, 100
Loewer, Barry, 26–28, 50n25, 87,
    88n14, 93
Lorentz, Hendrik, 30, 31, 38, 68

M

Mach, Ernst, 15, 67
Manifest image, viii, x, 43, 44,
    111–161
Many worlds, 48–50
Marmodoro, Anna, 78n6, 87n10
Marxism, vi
Matarese, Vera, 87n10, 88n13
Matter density field, 51–53
Maudlin, Tim, 4n2, 28, 37n19, 47,
    47n22, 51n29, 52n30, 55n33,
    59, 80, 91, 94
Maxwell, James Clerk, 30, 31, 68

McDowell, John, 123, 124
Measurement problem, 47, 49, 52,
    53, 60
Mele, Alfred R., 98n18, 99n19, 132
Mercati, Flavio, 11n9, 14n10, 44n21
Miller, Elizabeth, 82n7, 86
Millikan, Ruth Garrett, 152
Molière, 78
Mumford, Stephen, 78n6, 94
Myth of the given, 124, 129, 158,
    159

N
Naturalistic fallacy, vii, 121–123
Natural kinds, 148, 149
Necessity, 64, 89, 90, 99
Neuroscience, vi, 2, 66, 93, 117,
    119, 132, 157
Newton, Isaac, 1, 4, 9, 10, 12–15,
    29–31, 35, 58, 67
Newtonian mechanics, ix, 12, 16,
    18, 19, 23, 36–38, 43, 58
Non-locality, 55, 57–59
Normativity, x, 120–132, 140, 155,
    157, 159

O
O'Leary-Hawthrone, John, 147n16
Open society, vi

P
Panpsychism, 115n3, 138, 138n11
Papineau, David, 117n5
Past hypothesis, 26, 27, 57n35, 73,
    83–86

Perry, Zee, 82n7
Phase space, 23, 25
Physicalism, 120
Placement, 64, 65
Plato, v, vi, 133, 134
Popper, Karl, vi, 20n13, 136, 139
Powers, 78–80, 91, 94, 95, 106,
    159
Pragmatics, 127, 136
Price, Huw, 64
Primitive ontology, 5–9, 12, 14, 17,
    18, 21, 26, 29, 33, 39, 40, 42,
    44, 48, 49, 51, 53, 57, 60, 61,
    64–66, 69, 70, 74, 76–79, 81,
    82, 86–88, 90, 92, 99–101,
    104–106, 109, 111–113, 115,
    117–118, 121, 122, 130, 132,
    135–138, 156, 157, 160
Probability measure, 22, 23, 27, 50,
    60, 61
Process ontology, 119
Putnam, Hilary, 65n2
Pylkkänen, Paavo, 71n3, 88n12,
    115n4

Q
Qualia, 75, 114, 138
Quantum field theory, 59, 60
Quantum mechanics, 18, 44–62, 83,
    88, 93, 101, 107, 147n17

R
Reduction, 28, 72–74, 136, 138,
    145, 152
Reductionism, 72, 73, 145
Reichert, Paula, 23n14, 85n9

Relationalism, 12, 17, 28, 36, 39, 42, 80, 84
Relational mechanics, 11, 36, 44, 85
Res cogitans, 9, 111, 117, 140
Res extensa, 4, 8, 9, 111, 116, 117, 140
Rorty, Richard, 125, 151
Rovelli, Carlo, 32n16
Rule-following, 126, 129
Russell, Bertrand, 35, 36, 80

S

Sachse, Christian, 73
Sartre, Jean-Paul, 154, 154n20
Scanlon, T. M., 132n10, 142n12
Schnieder, Benjamin, 145n15
Schrödinger cat, 47, 48, 52
Schrödinger equation, 47, 48, 50, 101
Scientific image, vi–ix, 1–62, 64, 70, 74, 75, 77, 83, 85, 90, 111–123, 130–137, 139, 141, 149–156, 158–160
Scientific realism, viii, 43, 98, 125, 137, 149, 152, 154–156
Scientism, v, vi
Searle, John R., 120, 121n7
Seibt, Johanna, 115n3
Sellars, Wilfrid, viii, ix, 64, 115, 115n3, 116, 119, 124–126, 128–130, 136, 138, 140, 150–154, 158
Sellarsianism, 151–154, 151n18
Semantics, 63, 120, 121, 123, 126, 127, 135
Sense impressions, 124, 127, 139
Shape dynamics, 11, 36, 44, 59, 60, 83, 85
Simpson, William M. R., 87n10
Smith, Michael, 112

Special relativity, 38, 59
Spin, 53, 71, 81, 82, 88
Spinoza, 140, 141
Spontaneity, 154
Standard model, 8, 60, 71, 148
Statistical mechanics, 2, 24, 25, 44–62
Strawson, Galen, 92n15, 115n3
Structural realism, 70, 142
Struyve, Ward, 60n41
Super-determinism, 56
Super-Humeanism, 87–89, 91, 92, 100, 105, 106, 150, 158
Supervenience, 145, 146
Surplus structure, 10, 12, 14, 28, 31, 52, 79, 82, 109, 139
Swartz, Norman, 98n18
Symmetry, 7, 34, 79
Syntax, 120, 121, 123, 126
Systematicity, vii, 107

T

Teufel, Stefan, 49n24
Thales, 5
Thébault, Karim P. Y., 44n21
Tomassello, Michael, 143
Triangle inequality, 10, 80
Tumulka, Roderich, 48, 51, 51n28
Typicality, 23, 24, 27

U

Unification, 77, 77n5, 85, 91

V

van Brakel, Jaap, 131
van Inwagen, Peter, 95, 95n17, 98, 101, 102, 106

Vassallo, Antonio, 60n39
Velocity of light, 30, 36–38
Voltolini, Alberto, 154n19
von Wachter, Daniel, 95n16, 95n17

W
Wallace, David, 47n23, 48
Watson, James D., 66

Wave-function, 18, 19, 47–52,
    54, 55, 57–59, 71, 81,
    82, 86, 93, 100, 101,
    112
Weyl, Hermann, 42
Wheeler, John A., 34, 83
Wilson, Alastair, 87n10
Wittgenstein, Ludwig, 126,
    128

Printed in the United States
By Bookmasters